AF770267

BIOCHEMISTRY RESEARCH TRENDS

ORGANIC CHEMISTRY, BIOCHEMISTRY, BIOTECHNOLOGY AND RENEWABLE RESOURCES. RESEARCH AND DEVELOPMENT

VOLUME 1
TODAY AND TOMORROW

BIOCHEMISTRY RESEARCH TRENDS

Additional books in this series can be found on Nova's website under the Series tab.

Additional E-books in this series can be found on Nova's website under the E-book tab.

BIOTECHNOLOGY IN AGRICULTURE, INDUSTRY AND MEDICINE

Additional books in this series can be found on Nova's website under the Series tab.

Additional E-books in this series can be found on Nova's website under the E-book tab.

BIOCHEMISTRY RESEARCH TRENDS

ORGANIC CHEMISTRY, BIOCHEMISTRY, BIOTECHNOLOGY AND RENEWABLE RESOURCES. RESEARCH AND DEVELOPMENT

VOLUME 1
TODAY AND TOMORROW

GENNADY E. ZAIKOV
FRANK PUDEL
AND
GRZEGORZ SPYCHALSKI
EDITORS

New York

Copyright © 2013 by Nova Science Publishers, Inc.

All rights reserved. No part of this book may be reproduced, stored in a retrieval system or transmitted in any form or by any means: electronic, electrostatic, magnetic, tape, mechanical photocopying, recording or otherwise without the written permission of the Publisher.

For permission to use material from this book please contact us:
Telephone 631-231-7269; Fax 631-231-8175
Web Site: http://www.novapublishers.com

NOTICE TO THE READER

The Publisher has taken reasonable care in the preparation of this book, but makes no expressed or implied warranty of any kind and assumes no responsibility for any errors or omissions. No liability is assumed for incidental or consequential damages in connection with or arising out of information contained in this book. The Publisher shall not be liable for any special, consequential, or exemplary damages resulting, in whole or in part, from the readers' use of, or reliance upon, this material. Any parts of this book based on government reports are so indicated and copyright is claimed for those parts to the extent applicable to compilations of such works.

Independent verification should be sought for any data, advice or recommendations contained in this book. In addition, no responsibility is assumed by the publisher for any injury and/or damage to persons or property arising from any methods, products, instructions, ideas or otherwise contained in this publication.

This publication is designed to provide accurate and authoritative information with regard to the subject matter covered herein. It is sold with the clear understanding that the Publisher is not engaged in rendering legal or any other professional services. If legal or any other expert assistance is required, the services of a competent person should be sought. FROM A DECLARATION OF PARTICIPANTS JOINTLY ADOPTED BY A COMMITTEE OF THE AMERICAN BAR ASSOCIATION AND A COMMITTEE OF PUBLISHERS.

Additional color graphics may be available in the e-book version of this book.

Library of Congress Cataloging-in-Publication Data

ISSN: 2165-8676

ISBN: 978-1-62081-155-9

Published by Nova Science Publishers, Inc. † New York

Contents

Preface

"Not the one who has read a lot knows a lot
but the one who has comprehended a lot"
Kounfou

"The aim of Science is to lead our mind
following the Creator's reason"
Isaak Newton,
England

"I have not special talents
but I am very curious person"
Albert Einshtein,
Germany-Switzerland-USA

"Labour is Father of happiness"
Benjamin Franklin,
One from several founders of USA

"It is easy to see but it is difficult to foresee"
Benjamin Franklin

We included in this volume very important (and last) information in the field of organic chemistry, biochemistry, biotechnology, material science (synthesis, properties, applications) and renewable resources.

We did special emphases on the problems of microbiological corrosion of metals and protection against it, biodamaging and protection of paint materials, biodamages and protection of paper, biodegradation and protection of synthetic polymer materials, biodamages and protection of wood, coordination of specific retention volumes of organic compounds with their physical–chemical properties by means of polyparametric equations, renewable resources and plant biotechnology and hemp essential oil - production, characterization, application, enriched linen textiles as a material for "wellness" products.

Some part of volume is devoted to use of plant biotechnology methods for flax breeding in Latvia, exploring the potential of oilseeds as a sustainable source of oil and protein for aquaculture feed, potential and future prospects of dyeing plants: Madder *Rubia tinctorium*

L., Annato *Bixa orellana* L, Dyer's broom *Genista tinctoria* L., Woad *Isatis tinctoria*. L., genetic improvement of oat hulls and their potential for use as a renewable resource, application of the ultrasound treatment of flax fibers to produce yarns by open-end and wet flax spinning, preparation of rapeseed protein products for aquaculture feeding, sustainable biofuels - an international overview of existing approaches, Polish cultivar of white mulberry (*Morus alba* L.) and new directions of its utilization, evaluation of different metal oxide nano particle finishes, for self-cleaning of cotton fabric, biomass logistics: the key challenge of minimizing supply costs, ecological and economic optimisation of biogas plants.

Only fascination of Science is able to overcome natural disgust to the efforts of mind – all of us (Editors and contributors of this volume) were permanently deliberated these Isaak Newton's words when we have move along the reads of discoveries and understanding of the new idea in Organic Chemistry, Biochemistry, Biodegradation and Creation of New Materials on the base of Renewable Resources. These were the roads following the Creator's reason (please look for Epigrath of this volume above), the roads of acquisitions and losses, successful probes and humiliating errors, inspiration and despair, sacred delight and hopeless dejection. We were very happy to cooperate with contributors of this book.

In: Organic Chemistry, Biochemistry, Biotechnology ... ISBN: 978-1-62081-155-9
Editors: G. Zaikov, F. Pudel and G. Spychalski © 2013 Nova Science Publishers, Inc.

Chapter I

Use of Plant Biotechnology Methods for Flax Breeding in Latvia

***Dace Grauda*[1], *Andra Mikelsone*[2], *Aija Auziņa*[1], *Veneranda Stramkale*[3] *and Isaak Rashal*[1]**
[1]Institute of Biology, University of Latvia, Salaspils, Latvia
[2]Daugavpils University, Daugavpils, Latvia
[3]Agricultural Science Centre of Latgale, Viļāni, Viļānu distr., Latvia

Abstract

Two *in vitro* (anther and callus cultures) methods were applied for obtaining additional breeding initial material. Androgenic response of the used hybrids ranged from 1 to 11%, depending on genotype and growing conditions. Regeneration was only observed from diploid callus. A novel universal method of DNA fingerprinting was used to estimate genetic variability in somaclonal plants regenerants. The flax rust resistance alleles *L2, L6* and *L9* were detected by molecular markers in the breeding initial material.

Keywords: Flax, anther culture, callus culture, somaclonal variation, rust resistance, genetic variability

Introduction

Flax breeding in Latvia has an interrupted history. It was initiated in 1923 on the basis of local and Russian landraces. Six varieties had been bred in the time up to 1940, but after World War II flax breeding did not result in varieties and was cancelled in 1970. Contemporary flax breeding in Latvia was started again only 20 years ago, using the best repatriated Latvian flax genetic resources and advanced foreign varieties most applicable in Latvian climate conditions [1, 2].

Two *in vitro* methods (anther and callus cultures) were used in the Latvian flax breeding program. Anther culture enabled production of double haploid (DH) lines. DH lines are homozygous in all loci and therefore allow fast and efficient assessment of breeding material. It has been estimated that use of DH in flax breeding can reduce the time needed to obtain novel varieties by about 3-4 years, in comparison with traditional breeding [3, 4]. Somaclonal variation is an attribute of calli cultivation *in vitro* [5-7]. Three types of somaclonal variation have been described: heritable stable, heritable unstable and non-heritable (epigenetic) [8-10].

However, somaclonal variation occurs occasionally and its mechanism is not completely clear. Increased retrotransposon activity can be one of the sources of genetic changes during *in vitro* cultivation [11]. Somaclonal variation, which has been observed for many agronomic traits, such as disease resistance, date of heading and maturation, plant height, seed yield and quality etc. [12-15], has practical importance. However, the possibility to select and use somaclonal variants in breeding differs for specific traits and depending on the genotype of the particular variety. Somaclonal variation has been successfully used to derive new flax varieties and breeding lines with increasing resistance to biotic and abiotic stresses, plant height, number of seeds and number of seeds in a vessel [7, 16].

Flax resistance to microscopic fungi *Melampsora lini*, a causal agent of rust, is one of the important characteristics of breeding material quality. Flax rust occurs worldwide and can cause severe losses in seed yields and fibre quality. There are at least 34 genes conferring resistance to rust, which are classified into seven groups: K, L, M, N, P, D, and Q [17-21]. Several molecular markers can be used to detect the presence of a particular allele in flax accessions without direct infection tests, which are labor consuming and require maintaining a collection of pathogen isolates.

Aim and Background

The choice of appropriate varieties in crop productions is crucial for obtaining high-quality yield. An important task for Latvian flax breeding is to obtain high yielding, middle early varieties with good resistance to lodging and to flax diseases, and which are suitable both for conventional and organic farming. Breeding of flax, even nowadays, is a long and complicated process, which is based on hybridization and selection of the best plants. Taking in account facultative cross-pollination, the development of genetically stable lines can take more than 15 years using traditional methods [2, 3]. Therefore, the use of plant biotechnology (*in vitro*, molecular) methods in breeding process has immense potential use.

Table 1. Flax hybrids used for anther culture formation

Hybrid	Hybrid combination
E 37/2	Ceies / Norlin
E 37/49	Ceies / Norlin
E 38	Krupnosemjanij / Mikael

The aims of the study were to improve the anthers cultivation method for obtaining flax DH lines by optimization of embryo induction and regeneration conditions, to determine the optimal conditions for somatic callus induction and regeneration from local breeding material, to estimate change in activity of retrotransposons, and to evaluate Latvian flax genetic resources for the presence of rust resistance alleles.

Experimental Part

Anther Culture

Three flax F_4 hybrids and variety 'Lirina' as a control with known good androgenetic response [22] were used for the experiments (Table 1).

Donor plants were grown in a greenhouse. Flower buds with length 5.0-7.0 mm and width 1.5-3.0 mm were collected; pollen was collected at the middle or late uninucleate stage. Cold treatment (4 °C) was applied for 24 h. Buds were sterilized by dipping in a 50% solution of commercial bleach for 20 min, followed by rinsing three times with sterile water. Two different media were used for anther culture formation: 1) NC induction medium [4] with 6% sucrose, 0.4% agarose, and pH 5.5 and 2) NLN 82 medium [3] with 3% sucrose, 0.4% agarose, and pH 5.8. Petri dishes with anthers were incubated at 26 °C in the dark.

Each induced embryo that formed generative callus was transferred to one of three regeneration media: 1) NC regeneration medium [4] with 3% maltose, 0.8% gelrite, and pH 5.8, 2) MS/B5 medium (3% sucrose and 0.6% plant agar, pH 5.8) supplemented with 1 mg/l BAP, and 3) MS/B5 medium with addition of 1 mg/l BAP and 1 mg/l ascorbic acid.

Every four weeks generative calluses were subcultured onto fresh regeneration medium. During 16-54 weeks of cultivation, the ploidy level of generative calluses was determined using flow cytometry (Partec flow cytometer CyFlow Space).

Table 2. Accessions evaluated for flax rust resistance L alleles

Varieties	Landraces	Breeding lines
Ošupes 30	Blue di Riga	Rigaer 6/5
Ošupes 31	Riga Freis	Riga 27/12
Priekuļu 665	Riga Originario	Rigaer LIN 748/82
	Riga Vilnorin	Rigaer LIN 780/81
	Rigar B	N 101
	Vietējais 1	N 115
	Vietējais 3	N 143
	Vietējais 4	
	Vietējais 6	

Somatic Callus Culture

Two accessions were used for establishment of somatic callus cultures: 'Blue di Riga, a Latvian landrace with excellent adaptation to Latvian climatic conditions and with good seed and oil quality, and variety 'Lirina', with known good callus induction and regeneration capacity [28]. Flax seeds were sterilized by pre-treatment in 0.07% KMnO4 solution for 30 min, further treatment in 50% solution of commercial bleach (content of hypochlorite 4%) for 20 min, and final three times rinsing with sterile water. Sterilized seeds were germinated on Murashige and Skoog (MS) basal medium containing mineral salts and vitamins [23], 3% sucrose and 1% agar, supplemented by 20 mg/l AgNO3 (added after autoclaving), at pH 5.8. Cultures were incubated at 24 °C in light (intensity approx. 3000 lx) with a 16h/8h day/night photoperiod. Stem segments (2-4 mm) of *in vitro* grown seedlings (two leaf stage) were used as primary explants. All explants were placed on MS/B5 basal medium: MS mineral salts, B5 vitamins [24], 3% sucrose and 0.6% plant agar supplemented by 1 mg/l 2,4-dichlorophenoxyacetic acid (2,4 D) and 1 mg/l 6-benzylaminopurine (BAP, both added after autoclaving), pH 5.8. Cultures were grown in light (24 °C, 16h/8h day/night and light intensity approx. 3000 lx). After four weeks of cultivation, the calluses were placed on regeneration medium – MS/B5 medium supplemented by 1 mg/l BAP or 1 mg/l 2.4 D and 1 mg/l BAP. Cultures were grown in light (24 °C, 16h/8h day/night and light intensity approx. 3000 lx).

The universal IRAP (Inter-Retrotransposon Amplified Polymorphism) method [11] was used to detect possible retrotransposone movement. Genomic DNA was manually isolated using modified Shaghai-Maroof (1984) protocol. Content of the CTAB buffer was: 1M Tris, 0,5M EDTA, 10M NaCl, 2% CTAB, 2% PVP, at pH 8,0. DNA quality was tested by electrophoresis on 1% agarose gel for 1 hour, 70 V. PCR amplification was performed in 25 μl reactions on 96-well PCR plates. The cycling program was as follows: 3 min initial denaturation at 94 °C; 30 cycles of 95 °C denaturation (30 s), 50 °C primer annealing (40 s), and 68 °C extension (1 min); and a final extension for 10 min at 72 °C. The PCR products were resolved by electrophoresis for 15 h at 50 V on 1.7% agarose (Top Vision, Fermentas) in 1xTAE buffer followed by EtBr staining and UV visualization.

Thirty one (2074, 2076, 2077, 2078, 2079, 2080, 2081, 2083, 2220, 2232, 2237, 2238, 2239, 2242, 2270, 2271, 2272, 2273, 2373, 2374, 2375, 2376, 2377, 2378, 2383, 2385, 2386, 2389, 2390, 2394, 2415) primers [11] were evaluated for the potential use in flax breeding.

Three most polymorphic primers: 2077 (5' - 3' CTCACGATGCCA), 2080 (5' - 3' CAGACGGCGCCA), and 2271 (5' - 3' GGCTCGGATGCCA) were chosen for further analysis.

Detection of Flax Leaf Rust Resistance Alleles

The presence of *L2, L6* and *L9* rust resistance alleles in recently repatriated [1, 2] flax accessions of Latvian origin (Table 2) was tested. Genomic DNA from fresh leaves was extracted using the CTAB method [25]. Amplification of segments of the flax L locus was performed in a thermocycler. The 50 μL reaction volume contained 1xPCR buffer (10 mM Tris–HCl at pH 8.5, 50 mM KCl, 1.5 mM MgCl2, and 0.01% Tween-20), 250 μM of each dNTP, 25 ng of each primer, approximately 10-25 ng of genomic DNA, and 1 unit of Taq DNA polymerase (Fermentas). Following an initial 3 min denaturation step (93 °C), the reaction mixtures were subjected to 30 cycles of 1 min at 93 °C, 30 s at 55 °C, and 2 min at 72 °C. The last cycle was followed by an extension polymerization of 10 min at 72 °C.

Endonuclease digestions of PCR were performed using enzymes obtained from Fermentas following the recommendations of the manufacturer. PCR products were digested by *MboI* and the DNA restriction fragments were resolved on polyacrylamide gel. Polyacrylamide gel electrophoresis was carried out in TBE buffer, on 10% polyacrylamide gels at 100 V for 5 hours [17].

Results and Discussion

Anther Culture

All used genotypes formed embryos on the two media used (Figure 1).

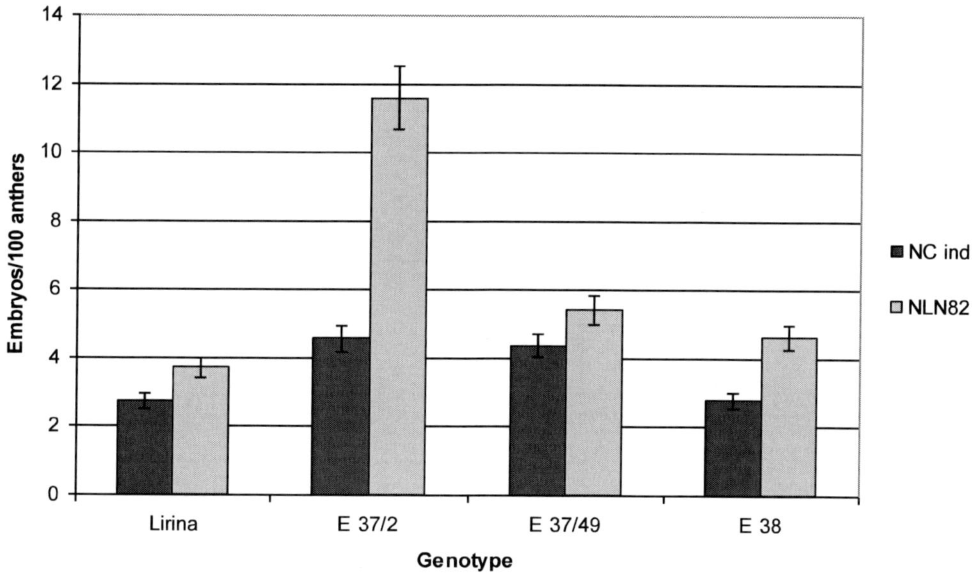

Figure 1. Embryo formation from anthers depending o induction medium.

Embryo yield (number of embryos per 100 anthers) ranged from 3 for the variety 'Lirina' and hybrid E38 on NC medium to 11 for the hybrid E37/2 on medium NLN 82. Frequency of embryo formation was higher on NLN 82 medium for all hybrids. The embryogenic response of hybrid E37/2 was about twice that of E37/49 on medium NLN 82, which is interesting because both originated from the same hybrid combination (Ceies/Norlin).

This difference possibly was due to lack of genetic uniformity of the parent varieties. Nevertheless, while hybrid E37/2 had the highest embryo yield (11%) on medium NLN 82, the yield was only 4% of anthers on the modified NC medium. The anther induction medium had effect not only on embryo formation, but also on generative callus formation and growth intensity (Figure 2). Calluses of all genotypes used grew better on the MS medium supplemented with vitamins from B5 medium (MS/B5). All calluses showed higher growth during cultivation on the NC induction medium, and addition of 1 mg/l ascorbic acid to

MS/B5 significantly increased growth of calluses on NLN82. Plants regenerants were obtained only from calluses formed from NLN82 embryos. The number of shoots from a callus ranged from 1 to 22.

Flow cytometry demonstrated that spontaneous diploidisation had occurred during flax anther culture. A total of 73% of calluses were mixsoploid (2n, 3n, 4n, 6n), most of which were 2n and 3n. Regeneration was observed only from diploid calluses.

The genotypes used had a intermediate rate of anther response (ranged from 3 to 11% embryo formation) comparing to yields obtained in other studies [3, 4, 22]. It has been previously found that factors such as genotype, composition of induction and regeneration media, and the transfer system of plant regenerants from *in vitro* to soil have crucial effect on plant regeneration [27, 28]. Presently, it is clear that the most sensitive stage is the callus induction media used for a particular genotype.

A

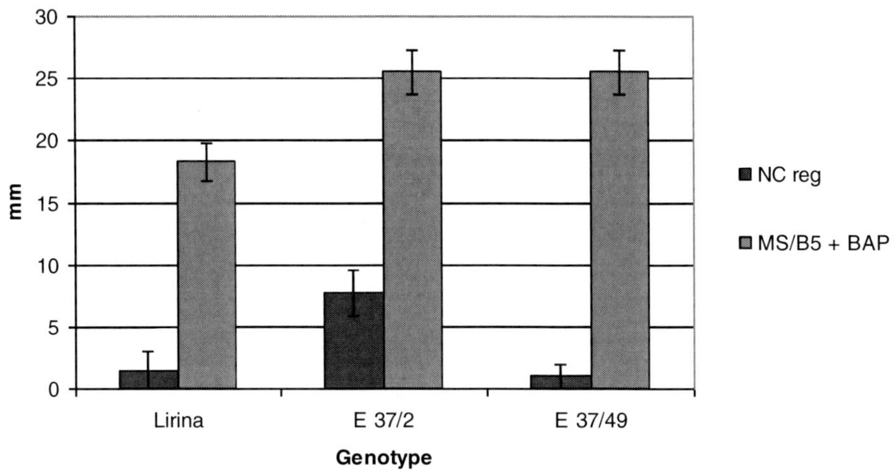

B

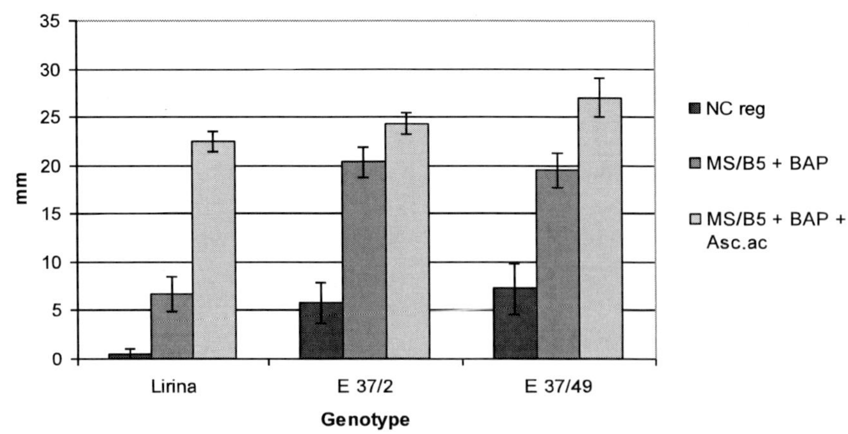

Figure 2. Callus diameter (mm) after 4 weeks cultivation on regeneration media: A) callus after cultivation on NC induction media, B) callus after cultivation on NLN82 induction media.

Somatic Callus Culture

Stem segments were previously found to be excellent for obtaining flax callus [27, 28]. In this experiment all planted stem segment explants formed callus on all induction mediums. Both plant genotype and phytohormone composition in the callus induction media had large effect on somatic callus growth and development (Figure 3). The correct choice of phytohormone balance enabled to obtain a higher proportion of callus that had regeneration zones: 10 to 80% for variety 'Lirina' and 75 to 95% for 'Blue de Riga'. The number of regeneration zones per callus varied from 3 to 40. Shoots of 3–4 cm length were cut and planted on the rooting medium.

Unfortunately, it is not possible to obtain a universal phytohormone combination for all flax genotypes [27, 28]. Therefore, to obtain a high number plants regenerants for breeding, it is necessary to experimentally find the optimal phytohormone composition to use in callus induction media for the sepcific material employed.

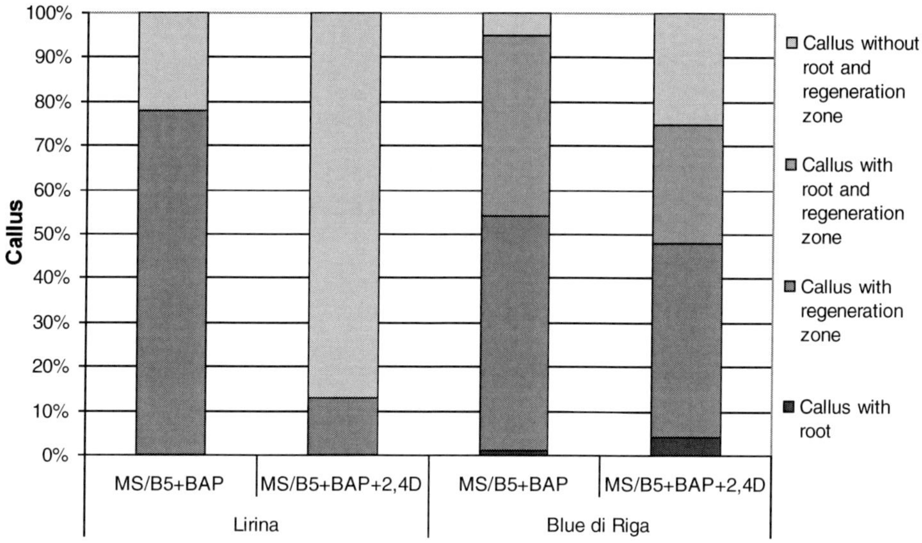

Figure 3. Stem segment callus after four weeks of cultivation on different callus induction media.

Plants regenerants from somatic callus culture were analyzed by the retrotransposon-based universal IRAP method [11]. A total of 72 loci were determined by the three primers used. New polymorphisms were detected among the plant regenerants (data are not shown), which indicated retrotransposons activity in flax callus culture.

Detection of Flax Leaf Rust Resistance Alleles

The leaf rust resistance alleles detected in the L loci are presented in Table 3. Flax rust resistance alleles *L2, L6* and *L9* were found in 9 of the tested 19 accessions. Four accessions of local origin had allele *L9*, four accessions had *L2*, and allele *L6* was detected in only one accession (N143). In Canada, resistance allele *L2* has been observed to be more effective [20], while in Russia allele *L6* had greater effect [21]. The results obtained suggest that accessions 'Blue di Riga', 'Rigaer LIN 780/81', 'Vietējais 3', 'Vietējais 6' and N143 could be useful as donors of rust resistance alleles. Investigation of the racial composition of the rust causal agent of Latvian population is necessary to specify resistance allele selection from the breeding point of view.

Table 3. Leaf rust resistance alleles in accessions of the Latvian origin

Genotype	Flax rust resistance alleles		
	L2	*L6*	*L9*
Blue di Riga	+		
Riga Freis			+
Rigar B			+
Rigaer 6/5			+
Rigaer LIN 780/81	+		
Vietējais 3	+		
Vietējais 4			+
Vietējais 6	+		
N 143		+	

Conclusion

The experiments on anther culture formation showed that all of the used genotypes had an intermediate anther response. Nevertheless, correct choice of the most appropriate phytohormone composition in the cultivation medium can increase frequency of embryo formation. It is important is to find the suitable anther induction medium for the particular genotype employed. For the investigated hybrids, the best medium was NLN 82. Flow cytometry demonstrated that spontaneous diploidisation had occurred during flax anther culture. Regeneration was observed only from diploid gametic calli.

All of the used varieties were excellent for callus culture formation (100% of used explants formed callus). Callus regeneration capacity can be increased by using the optimal phytohormone composition in callus induction medium. Shoots were formed by all callus with regeneration zones. Tests with retrotransposone based markers showed that callus cultivation is related to increased retrotransposons activity. Nine accessions possess resistance *L* alleles, among them 5 had effective alleles *L2* or *L6*. These accessions are recommended as source material of rust resistance for flax breeding.

References

[1] Rashal I., Stramkale V. (1998) Conservation and use of the Latvian flax genetic resources. Proceedings of the Symposium "Bast Fibrous Plants Today and Tomorrow. Breeding, Molecular Biology and Biotechnology Beyond 21th Century", 28-30 September 1998, St. Petersburg, Russia. *Natural Fibres,* 2, p. 56-58.

[2] Grauda D., Stramkale V., Rašals I. (2004) Evaluation of Latvian flax varieties and hybrids. *Proceedings in Agronomy*, 6, p. 159-165.

[3] Bergmann, R; Friedt, W. (1997) Haploidy and related biotechnological methods in linseed (*Linum usitatissimum* L.). In: *In vitro Haploid Production in Higher Plants.* Jain M.S., Sopory S.K., Veilleux R.E. (eds). Kluwer Academic Publishers, Netherlands, Dordrecht, p. 1-9.

[4] Nitchterlein, K. (2003) Anther culture of linseed (*Linum usitatissimum* L.). In: *Doubled Haploid Production in Crop Plants. Maluszynski* M., Kasha K.J., Foster B.P., Szarejko I. (eds.), Kluwer Academic Publishers, Netherlands, Dordrecht, p. 249-254.

[5] Larkin, P.J.; Scowcroft, W.R. (1981) Somaclonal variation – a novel source of variability from cell culture for plant improvement. *Theor Appl Genet,* 60, p. 197-214.

[6] Leike, H. (1985) Methoden der Gewebekultur zur Beschleunigung des Zuchtfortschrittes. *Tagungoberichts Akademie Landwirtschaft – Wissenschaft.,* 237, S. 19-33.

[7] Rutkowska-Krauze, I, Mankowska, G, Lukaszewicz, M, Szopa J. (2003) Regeneration of flax (*Linum usitatissimum* L.) plants from anther culture and somatic tissue with increased resistance to Fusarium oxysporum. *Plant Cell Rep*, 22, p.110-116.

[8] Karp A. (1991) On the current understanding of somaclonal variation. In: *Oxford Surveys of Plant Molecular and Cell Biology.* B.J. Miflin (eds.),7, p. 1-58.

[9] Vázquez A.M. (2001) Insight into somaclonal variation. *Plant Biosystems,* 135, p. 57-62

[10] Bednarek P., Orłowska R., Koebner R., Zimny J. (2007) Quantification of the tissue-culture induced variation in barley (*Hordeum vulgare* L.). *BMC Plant Biol.,* 7:10, doi:10.1186/1471-2229-7-1.

[11] Kalendar R., Antonius K., Smýkal P., Schuman A. H. (2010) iPBS: a universal method for DNA fingerprinting and retrotransposon isolation. *Theor Appl Genet*, 12, p. 1419-1430.

[12] Ahloowalia B.S. (1987) Plant regeneration from embryo-callus culture in barley. *Euphytica,* 36, p. 659-665.

[13] Choo T.M., Li J., Ho K.M., Kong D., Narisimhalu P. (1992) Somaclonal variation in Leger barley. *Proceedings of the SABRAO International Symposium on the Impact of Biological Research on Agricultural Productivity,* SABRAO, p. 163-174.

[14] Karp A. (1994) Origins, causes and uses of variation in plant tissue cultures. In: *Plant Cell and Tissue Culture.* Vasil I.K., Thorpe T.A (eds.), Kluwer Academic Publishers, Dordrecht, p. 139-151.

[15] Bregitzer P., Poulson M. (1995) Agronomic performance of barley lines derived from tissue culture. *Crop Sci.*, 35, p. 1144-1148.

[16] Poliakov, A.V. (2000) *Biotechnology in flax Breeding.* Tver, Russia, Format 84 p. (in Russian).

[17] Hausner G., Rashid K. Y., Kenaschuk E. O., Procunier J. D. (1999) *The development of codominant PCR/RFLP based markers for the flax rust-resistance alleles at the L locus.* Genome, 42, p. 1-8.

[18] Luck J., Lawrence G., Dodds P, Shepherd K., Ellis J. (2000) Regions outside of the leucine-rich repeats of flax rust resistance proteins play a role in specificity determination. *Plant Cell*, 12, p. 1367-1378.

[19] Anderson P., Lawrence G., Morrish B., Ayliffe M., Finnegan E., Ellis J. (1997) Inactivation of the flax rust resistance gene m associated with loss of a repeated unit within the leucine-rich repeat coding region. *The Plant Cell,* 9, p. 641-651.

[20] Islam M., Shepherd K., Present status of genetics of rust resistance in flax. (1991) *Euphytica*, 55, p. 255-267.

[21] Kutuzova S. N. (1994) Genetic bases of long-term resistance to rust in flax varieties. *Genetika,* 30, p. 1363-1373 (in Russian).

[22] Burbulis N., Blinstrubiene A. (2006) Comparison of anther culture response among *Linum usitiatissimum* L. cultivars and their hybrids. *Acta Universitatis Latviensis, Biology,* 701, p. 131-138.

[23] Murashige T., Skoog F. A (1962) revised medium for rapid growth and bioassays with tobacco tissue cultures. *Physiol. Plant,* 15, p. 473-497.

[24] Gamborg O. L., Miller R. A., Ojima K. (1968) Nutrient requirements of suspension cultures of soybean root cells. *Exp Cell Res* 50, p.151-158.

[25] Cota-Sanchez J. H., Remarchuk K., Ubayasena K. (2006) Ready-to-use DNA extracted with CTAB method adapted for herbarium specimens and mucilaginous plant tissue. *Plant Mol Biol Rep* 24, p.161-167.

[26] Saghai-Maroof M.A., Soliman K.M., Jorgensen R.A., Allard R.W. I. (1984) Ribosomal DNA spacer-length polymorphisms in barley: Mendelian inheritance, chromosomal locations and population dynamics. *Proc. Natl. Acad. Sci.*, USA, 81, p. 8014-8018

[27] Grauda D., Rashal I., Stramkale V. (2006) The use *in vitro* methods for obtaining flax breeding source material. In: "*Renewable Resources and Plant Biotechnology",* Eds. R. Kozlowski, G. E. Zaikov, F. Pudel. Nova Publishers, New York, p. 127-134.

[28] Grauda D., Mikelsone A., Rashals I. (2009) Use of antioxidants for enhancing flax multiplication rate in tissue culture. *Acta Horticulture,* 812, p. 147-151.

In: Organic Chemistry, Biochemistry, Biotechnology ... ISBN: 978-1-62081-155-9
Editors: G. Zaikov, F. Pudel and G. Spychalski © 2013 Nova Science Publishers, Inc.

Chapter II

Exploring the Potential of Oilseeds as a Sustainable Source of Oil and Protein for Aquaculture Feed

***Crystal L. Snyder*[1], *Paul P. Kolodziejczyk*[1,2], *Xiao Qiu*[3], *Saleh Shah*[4], *E. Chris Kazala*[1] *and Randall J. Weselake*[1]**

[1]Department of Agricultural, Food and Nutritional Science,
University of Alberta, Edmonton, Alberta, Canada
[2]Biolink Consultancy Inc., New Denver, B.C., Canada
[3]Department of Food and Bioproduct Sciences,
University of Saskatchewan, Saskatoon, Saskatchewan, Canada
[4]Alberta Innovates-Technology Futures, Vegreville, Alberta, Canada

Abstract

Aquaculture is currently the world's fastest growing primary food production sector, but it is also a leading consumer of fish oil and fish meal from capture fisheries, which is a major concern for the industry's long-term sustainability. Oilseeds have been widely explored as an alternative land-based source of oil and protein for aquaculture feeds, but their use has been limited because conventional seed oils do not contain very long chain ω-3 fatty acids such as eicosapentaenoic acid (EPA) and docosahexaenoic acid (DHA). Over the past decade, however, there has been considerable progress toward the genetic engineering of oilseeds to produce very long chain ω-3 fatty acids. Such modified oilseeds show promise as nutritional supplements for aquafeed applications. Advances in oilseed processing technology may also assist in the development of sustainable oilseed-based aquafeeds.

Keywords: Aquaculture, aquafeed, nutrition, fish oil, ω-3 fatty acids, polyunsaturated fatty acids, protein, oilseeds

Introduction

Aquaculture currently supplies almost half of the seafood for human and animal consumption [1], and as one of the world's fastest growing food production sectors, the industry is under pressure to address the sustainability of its supply chain, particularly with respect to its reliance on fish oil and fish meal from capture fisheries [2]. The growing concern over the decline of the world's capture fisheries has prompted considerable research into alternative aquafeed ingredients, especially those from land-based plants. Oilseeds, for example, have been widely explored as an alternative source of oil and protein in fish diets, but present a number of challenges that currently limit their use in aquafeeds. Unlike marine oils, which are enriched in very long chain (20-22 carbon) polyunsaturated fatty acids (VLCPUFA) such as eicosapentaenoic acid (EPA, 20:5ω-3) and docosahexaenoic acid (DHA, 22:6ω-3), conventional seed oils are rich in 18-carbon fatty acids (mainly linoleic acid, 18:2ω-6 and α-linolenic acid, 18:3ω-3), which fish are unable to efficiently convert to VLCPUFA. In addition, animals lack the ability to convert omega-6 fatty acids to omega-3 fatty acids, making most high linoleic vegetable oils an inadequate substitute for fish oil. Maintaining high levels of ω-3 VLCPUFA in farmed fish is of paramount concern not only for optimal fish health, but also for the nutritional quality of the end products destined for human consumption. Indeed, the relatively low consumption of ω-3 VLCPUFA in the Western diet, together with high intake of ω-6 fatty acids, has been blamed in part for the growth of chronic, diet-related disease among Western populations [3].

While oilseeds offer many advantages for the sustainable production of aquafeed ingredients, engineering oilseeds to produce VLCPUFA has proven to be a formidable biotechnological challenge. Over the past decade, research groups around the world have explored various strategies for producing some or all of the fatty acids in the VLCPUFA biosynthetic pathway [4]. Almost all of these approaches require the insertion and coordinated expression of multiple transgenes, the effectiveness of which tends to vary depending on the nature of the genetic construct and of the host plant [5,6]. To date, substantial accumulations of various precursors leading to the formation of arachidonic acid (ARA, 20:4ω-6) and EPA have been obtained [4]; however, commercially-viable levels of DHA have yet to be achieved.

Although much of the commercial value of commodity oilseeds is in the oil fraction, the meal of many oilseeds, if processed appropriately, can be used as a high-quality protein source for animal feeds. Research is ongoing into the use and enhancement of oilseed meal for aquafeed, with many studies focusing on the enrichment of high-value bioactive components (e.g. carotenoids) or reduction of anti-nutritional factors [7]. At the same time, advances in oilseed processing technology are also helping to maximize the value of the oil and the meal for aquafeed and other applications.

This paper will review progress toward the production of ω-3 fatty acid-enriched seed oils for aquaculture feeds, their efficacy as aquafeed ingredients, and the role of processing technologies in maximizing the potential of oilseeds for aquaculture applications.

Engineering Oilseeds for the Sustainable Production of ω-3 Fatty Acids for Aquaculture

In animals, linoleic acid (LA, 18:2ω-6) and α-linolenic acid (ALA, 18:3ω-3) are considered to be the only dietary fatty acids that are essential for the production of VLCPUFA, including EPA and DHA. Animals possess a biosynthetic pathway for the further desaturation and elongation of linoleic and α-linolenic acids to form ARA and EPA, respectively (Figure 1A). However, the poor efficiency of this process, particularly the initial Δ6 desaturation step, can severely limit the accumulation of VLCPUFA in many individuals [8]. As a result, direct consumption of dietary VLCPUFA is recommended for optimal nutrition [9].

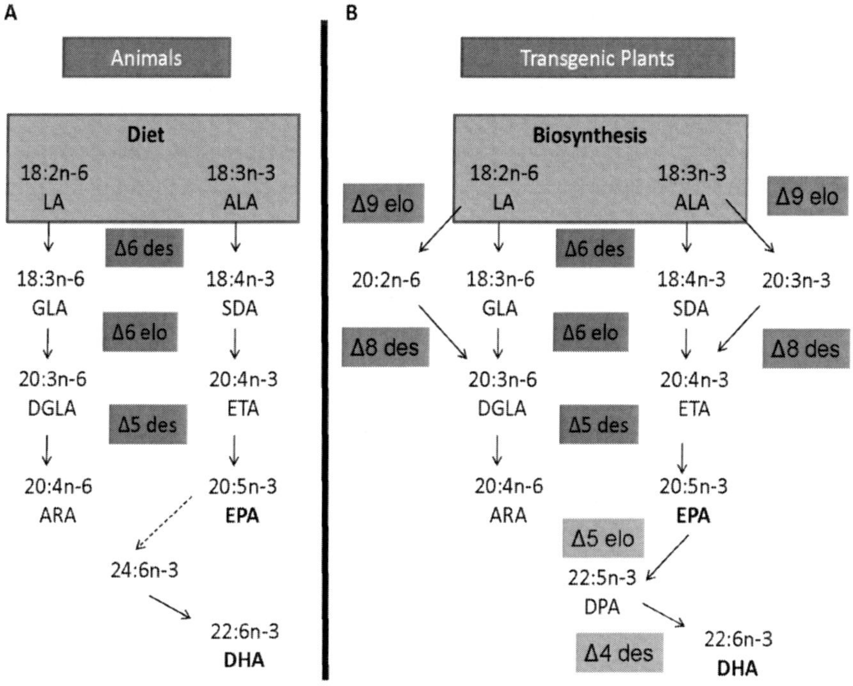

Figure 1. Pathways for very long chain polyunsaturated fatty acid (VLCPUFA) biosynthesis in animals and plants. A) In vertebrates, VLCPUFA are synthesized from the dietary essential fatty acids, linoleic acid (LA) and α-linolenic acid (ALA), by the Δ6-desaturation pathway, consisting of a Δ6-desaturase Δ6-elongase, and a Δ5-desaturase, leading to the formation of arachidonic acid (ARA; ω-6 pathway) and eicosapentaenoic acid (EPA; ω-3 pathway). Docosahexaenoic acid (DHA) is synthesized from EPA following further elongation and desaturation to 24:6, which undergoes one round of β-oxidation to form DHA. In most vertebrates, however, the Δ6-desaturation step is limiting, resulting in poor rates of conversion to EPA and DHA. B) Two desaturation/elongation pathways have been tested extensively in transgenic plants. The Δ6-desaturation pathway (red) is analogous to the pathway present in animals, but uses LA and ALA generated through *de novo* biosynthesis. The Δ9-elongation pathway (blue) also leads to the formation of ARA and EPA, but elongation takes place first, and is followed by two desaturation steps. From EPA, at least two additional activities (purple) are required to generate DHA. (des: desaturase; elo: elongase. All fatty acid abbreviations are defined in the text. The notation n-3/n-6 and ω-3/ω-6 are used interchangeably.)

Even fish, which are generally considered to be excellent dietary sources of VLCPUFA for humans, must obtain their VLCPUFA from their diet, usually through bioaccumulation of VLCPUFA produced by marine algae and other microorganisms.

The VLCPUFA-producing pathways of these marine microorganisms have offered a diverse array of genetic tools for oilseed biotechnology. Unlike animals, these microorganisms possess multiple pathways for VLCPUFA biosynthesis. Some organisms rely on aerobic desaturation/elongation processes analogous to the pathway present in animals [10], while others use a polyketide synthase (PKS) to build VLCPUFA in a process similar to *de novo* fatty acid biosynthesis in plants [11]. A few higher plants, including borage and echium [12,13], also possess the critical Δ6 desaturation activity and naturally accumulate gamma-linolenic acid (GLA, 18:3ω-6) and stearidonic acid (SDA, 18:4ω-3); however, these plants lack the enzyme activities required to convert GLA or SDA to downstream VLCPUFA. Nevertheless, these plants have served as an important source of genes supporting oilseed biotechnology efforts.

Indeed, the Δ6 desaturation pathway has been one of the most extensively studied in transgenic plants. Expression of recombinant desaturases from various sources has resulted in relatively high accumulation of GLA, and to a lesser extent, SDA. In *Brassica juncea*, expression of a Δ6 desaturase alone yielded up to 40% GLA and 8% SDA [14]. Other groups have used additional Δ12 or Δ15 desaturases to increase the supply of LA or ALA precursors, respectively [15,16]. Using a high linoleic safflower (*Carthamus tinctorius*) line coexpressing a Δ6 and Δ12 desaturase, Nykiforuk et al, [5] were able to obtain more than 70% GLA in the seed oil; these lines have recently been approved for commercialization as SonovaTM 400. Achieving high levels of SDA, particularly without significant coproduction of GLA, has been somewhat more challenging. To this end, an ω-3-selective Δ6 desaturase was introduced into flaxseed in order to preferentially increase the accumulation of SDA [17]. SDA-enriched seed oils, having bypassed the limiting Δ6 desaturation step, may be particularly useful for aquafeed applications; however, the efficacy of these oils is still being tested in fish [13,18].

Conversion of GLA and SDA to ARA and EPA, respectively, requires the introduction of at least two additional genes encoding a Δ6 elongase and a Δ5 desaturase to complete the "conventional" Δ6 desaturation pathway (Figure 1B). This pathway was first reconstituted in tobacco (high LA background) and flax (high ALA background), resulting in approximately 30% GLA or SDA, respectively, but only up to about 5% total 20-carbon VLCPUFA [19]. It appeared that the Δ6 elongation step was not very efficient, even though the recombinant elongase was successfully expressed in developing seeds and exhibited *in vitro* enzyme activity. Analysis of the acyl-CoA pool of transgenic flaxseed, however, indicated that there were only trace levels of Δ6 desaturated PUFA in the acyl-CoA pool, where elongation occurs [19]. Since the Δ6 desaturase utilizes phospholipid substrates and the elongase requires acyl-CoA substrates [20], this so-called "substrate dichotomy" was suggested to be a major limiting factor for VLCPUFA formation in transgenic plants [19]. These observations were supported by another study [21] employing an "alternative" Δ9 elongation pathway (Figure 1B), in which LA and ALA are first elongated, and then desaturated twice to form ARA and EPA. In this pathway, there is no need to transfer of substrates from phospholipids to the acyl-CoA pool and then back again. When this pathway was introduced into *Arabidopsis* leaves, up to 7% ARA and 3% EPA was obtained, and the total levels of 20-carbon PUFA were greater than 20% [21].

Other groups have tested a number of variations on the Δ6 desaturation pathway in efforts to overcome this problem of substrate dichotomy. Adding a Δ12 desaturase and an additional Δ6 elongase to the minimal three gene pathway resulted in increased elongation efficiency in transgenic *Brassica juncea* seeds, yielding up to 25% ARA [22]. The further introduction of an ω-3 desaturase, which converts ω-6 intermediates to their ω-3 counterparts, resulted in up to 11% EPA [22]. Zero erucic acid *Brassica carinata*, however, seems to be a more suitable host for EPA production, accumulating up to 20% EPA in lines expressing a five gene construct (Δ12 desaturase, Δ6 desaturase, Δ6 elongase, Δ5 desaturase, and an ω-3 desaturase) [6]. In zero-erucic acid *B. juncea*, the same construct yielded only 5% EPA[6], which highlights the importance of host species on the outcome of such transgenic experiments.

Some groups have also explored the use of acyl-CoA desaturases as another approach to overcoming substrate dichotomy in transgenic plants. In this strategy, Δ6 desaturation and elongation both occur in the acyl-CoA pool. Transgenic soybeans expressing an acyl-CoA/phospholipid desaturase from liverwort accumulated more than 19% total 20-carbon PUFA [23]. Although this increase comprised mostly ω-6 fatty acids, reflecting the high levels of linoleic acid precursor in soybean, this study demonstrated the effectiveness of acyl-CoA desaturases in overcoming substrate dichotomy. Another study using an ω-3-specific acyl-CoA desaturase showed a similar proof-of-concept in *Arabidopsis*, but achieved much lower total levels of VLCPUFA [24].

While it appears that the nature of the genetic construct, source of the transgenes, and selection of the host species all play a crucial role in determining the success of VLCPUFA-engineering experiments, recent advancements in our basic understanding of plant lipid biochemistry are also shedding insights into the biochemical basis for substrate dichotomy. The discovery of new enzymes involved in substrate trafficking [25,26] and the elucidation of a proposed "acyl-editing" pathway in developing oilseeds [27,28] will greatly assist ongoing efforts to develop a sustainable land-based source of VLCPUFA.

Although much of the research to date has focused on EPA as an essential first step in achieving high levels of VLCPUFA in transgenic plants, it is well recognized that DHA is equally important for aquaculture applications Several studies have shown proof-of-concept production of DHA in transgenic systems. Production of DHA has been technically more difficult, both in terms of overcoming substrate dichotomy and in handling the very large multi-gene constructs required for DHA production. Wu et al [22], for example, used a nine-gene construct to demonstrate proof-of-concept DHA synthesis in *Brassica juncea.* Recent methodological advancements such as the transient leaf-based assay described by Wood et al[29], promise to accelerate development of complex multi-gene constructs by facilitating rapid testing and interchangeability of independent genetic elements. This technique has already been successfully applied to the VLCPUFA pathway, resulting in up to 2.5% DHA in tobacco leaves and similar levels in stably-transformed *Arabidopsis* seeds [30]. Given the rapid pace of research in this area, it is likely only a matter to time before commercially viable levels of DHA are achieved in a crop plant [31-34].

Efficacy of ω-3 Vegetable Oils in Fish Nutrition

Over the past few decades, a wide range of vegetable oils and vegetable oil blends have been explored as possible alternatives to fish oil for aquafeed applications (see [35,36]for recent reviews). All vertebrates, including fish, have certain dietary requirements for essential fatty acids (EFA). In fish, dietary EFA requirements differ according to species; freshwater and diadromous fish may only require 18-carbon PUFA (eg. LA or ALA), while marine species require ω-3 VLCPUFA (eg. EPA and DHA) [13]. A deficiency in EFA can lead to adverse effects on growth and reproduction and in severe cases, could result in increased mortality [37]. The diversity of vegetable oils available and the differences in the dietary needs of cultivated fish species have led to a variety of different strategies for the incorporation of vegetable oils into aquaculture feeds. Some of these studies will be reviewed here.

Although flaxseed is a relatively minor commodity oilseed in terms of total production, it has received a great deal of attention for aquafeed applications due to its high levels of ALA (50-60%). One of the earliest fish oil replacement studies compared flax oil, rendered animal fat, and salmon oil in the dry diet of Chinook salmon (*Onchorhynchous tshawytscha*) over sixteen weeks [38]. Fish growth was not affected by the substitution, but the flesh fatty acid profiles, particularly PUFA content, varied according to the fatty acid composition of the diet. Several other fish oil replacement studies have reported similar trends; that is, the tissue fatty acid composition largely reflects that of the diet, but vegetable oils generally do not impair growth or performance of the fish [39-41]. After comparing two base diets (100% fish oil or 100% flax oil) substituted with various proportions of sunflower oil, Menoyo et al [40], concluded that flax could be a viable source of dietary lipids for farmed Atlantic salmon. In trout (*Onchorhynchus mykiss*), studies comparing salmon oil, flax oil, soybean oil, poultry fat, pork lard and beef tallow suggested that the EPA and DHA levels in the tissue were regulated and maintained at 10-12% regardless of diet [41].

As a Δ6 desaturated ω-3 PUFA, stearidonic acid (SDA) is also particularly interesting, not only for aquaculture, but for human and livestock nutrition as well. In mammals, the conversion rate of SDA to EPA is around 17-33% [42-44], while the conversion of ALA to EPA is much lower at 0.2-7% [45]. The low rate of conversion of ALA to EPA is believed to be due to the inefficient Δ6 desaturation step; thus, consumption of SDA could effectively bypass this step and support more efficient accumulation of downstream VLCPUFA. As fish utilize a similar pathway for converting ALA to EPA, SDA-enriched oils may be useful for maintaining or increasing EPA levels in the tissue.

To date, most studies investigating the efficacy of SDA have used echium oil as a source of SDA. Echium is one of a few higher plants that accumulate appreciable amounts of SDA (up to 14%). In Arctic charr (*Salvelinus alpines*), replacing 80% of the dietary fish oil with echium oil for 16 weeks led to increases in dihomo-gamma-linolenic acid (DGLA, 20:3ω-6) and eicosatetraenoic acid (ETA, 20:3ω-3), the immediate elongation products of GLA and SDA, respectively, in both flesh and liver tissue [46]. The echium diet, however, was not effective at maintaining EPA and DHA levels in charr. In a similar trial with Atlantic cod (*Gadus morhua*), a 100% echium oil diet for 18 weeks resulted in only slight increases in DGLA, with no increase in ETA [18]. Such studies illustrate the complexity of species-specific differences in dietary fatty acid metabolism.

In Atlantic salmon parr, neither echium oil nor a 1:1 blend of echium and canola oil resulted in a decrease in white muscle DHA content after 42 days [47]. It was suggested that reduced availability of VLCPUFA in the diet led to efficient retention and redistribution of DHA [48]. More recent studies with echium oil in salmon [49] and barramundi [50] failed to show any major benefits of SDA-enriched oils in fish feeding trials. It is worthwhile noting, however, that the SDA content of echium (14%) is relatively low compared to what may be obtainable from a transgenic plant source, and the presence of GLA as a competing ω-6 byproduct may also affect the conversion efficiencies observed in feeding trials. Thus, a plant source high in SDA and low in GLA and other ω-6 PUFA may be more effective at shifting fish metabolism toward conversion of ω-3 PUFA.

Given that the flesh fatty acid composition typically reflects the lipid composition of the diet, it becomes clear that direct substitution of fish oil with EPA- and DHA-enriched oils may be the most effective means of meeting the ω-3 VLCPUFA requirements of farmed fish. As described earlier, efforts to produce a sustainable land-based source of EPA and DHA are advancing rapidly; however, the intensifying pressure on the global fish oil supply requires a multi-faceted approach to address the sustainability of the aquaculture sector. To that end, the use of different feeding phases in fish production has been explored as a way to make the most efficient use of available dietary lipids (reviewed by [35]). These trials use different starter, grow-out, and finishing feeds to address the different dietary needs at various stages of development. With respect to fish development, the need for dietary VLCPUFA (i.e. fish oil) is greatest during the starter phase, in which rapid growth, high survival, and normal development (particularly of neural tissue) are the most important considerations. Finishing feeds, though not yet common in commercial settings, may also call for a greater VLCPUFA content in order to ensure that the desired post-harvest fatty acid composition is obtained.

There seems to be the greatest potential for fish oil substitution during the grow-out phase, during which the flesh fatty acid composition is not as critical and growth rate and feed conversion are the main considerations. In this case, a VLCPUFA-rich finishing diet would facilitate "washing-out" of the less desirable flesh FA prior to harvest. Such feeding strategies would allow fish producers to reduce their overall reliance on fish oil through more efficient feeding strategies, thereby reducing feed costs and enhancing the sustainability of the operation.

Oilseed Processing Technologies for Aquafeed

There has been considerable progress toward developing sustainable alternatives to fish meal and fish oil for aquafeed applications [51]. Today, the pork, beef, and poultry industries are the major users of oilseed meals for feed applications, but advancements in processing technology may expand market opportunities into the aquaculture sector. While the development of specialty oils containing high levels of physiologically important PUFA remains a major focus, advancements in oilseed processing technologies are also crucial for the production of high-quality protein concentrates for fish nutrition. For example, plant seeds contain a substantial amount of fibre, which is considered undesirable for fish feed. Novel processes based on fractionation of the oilseed cake or meal may facilitate the removal of anti-nutritional components left behind by traditional processing [52] and yield a product more amenable for use in fish feed.

The interest in processing technologies has been driven in part by the growth of so-called biorefineries, which aim to maximize the value of all components of a particular feedstock. The rapid expansion of oilseed-based biofuels has sparked increased interest in finding alternative markets for the leftover seed meal, which would help offset the relatively high feedstock costs involved in biodiesel production. Technologies that recover high-value components from the meal or otherwise increase the value of the meal for feed applications would substantially broaden the operating margins for biorefining facilities while helping to meet the needs of specialty markets such as the aquaculture sector.

The typical industrial process for production of edible oils involves pressing the seeds, followed by a solvent extraction, usually with hexane. This method results in a good recovery of high-quality, low cost oil, with residual seed meal being used by the animal feed industry. However, environmental considerations, as well as the desire to increase the value of the meal, have led researchers to explore processing technologies that rely less on organic solvents and take advantage of aqueous or enzyme-assisted technologies. Enzymes can be used to facilitate oil extraction by hydrolyzing the cell wall polysaccharides and oilbody membranes, and have the advantage of yielding non-denatured protein, which has higher value as a protein concentrate (containing fibre) or protein isolate (fibre removed). Advancements in industrial enzyme production are making enzyme-assisted processes increasingly cost-efficient, and several modifications of aqueous processes have been described in papers and patents [52,53]. Several companies have already patented technologies for production of canola and flax protein isolates, and industrial pilot plants using aqueous technologies are already in operation.

Several processes for the manufacture of protein concentrates or isolates for aquaculture have also been developed. These technologies yield non-denatured proteins with better digestibility for fish nutrition. Membrane separation techniques are used to separate oilseed components, remove anti-nutritional factors, and fractionate protein for fish feed production.

VLCPUFA-enriched oilseeds pose additional challenges for processing since the high PUFA content of the oils reduces their thermal and oxidative stability. Aqueous milling processes, followed by separation of the lipid fraction, are well-suited for the preservation of the high-value VLCPUFA, since the harsh refining steps (bleaching, degumming and deodorization) are not needed for fish feed production. The presence of phospholipids is actually considered beneficial for fish nutrition.

Supercritical fluid extraction is another possible strategy for processing of oilseeds for aquafeed applications, but so far the expansion of this technology has been inhibited by the cost of the process and lack of equipment for continuous processing using this technology [54].

Conclusion

Aquaculture production has grown immensely since the mid-1960s, and has now become the world's fastest growing primary food production sector, providing almost half of all the fish consumed by humans today [55]. At the same time, production from capture fisheries has declined due to chronic overfishing – largely to support the expansion of the global aquaculture industry. In 2006, global aquaculture utilized 68% of fish meal and 88% of fish oil produced worldwide [56], with a few carnivorous species (e.g. trout and salmon) consuming 19.5% and 51% of meal and oil, respectively, despite the fact that they represent only 3% of global aquaculture production [56].

Clearly, neither capture fisheries nor aquaculture can sustainably meet the needs of a human population expected to exceed 9 billion by 2050, unless the aquaculture industry can overcome its reliance on capture fisheries as a source of feed ingredients. Future growth of the industry depends on the development of alternative feed ingredients, a large fraction of which are likely to come from terrestrial plants, particularly from modified oilseeds containing EPA and DHA. Advancements in oilseed biotechnology and processing have set the stage for such designer oilseeds to make a major impact on the aquafeed market in the years to come.

Acknowledgments

All co-authors acknowledge the support from AVAC Ltd., and RJW thanks the Canada Research Chairs program.

References

[1] Subasinghe R, Soto D, Jia J: Global aquaculture and its role in sustainable development. *Rev Aquaculture* 2009, 1: 2-9.

[2] de Silva S, Francis D, Tacon A: Fish oils in aquaculture. In *Fish Oil Replacement and Alternative Lipid Sources in Aquaculture Feeds*. CRC Press; 2010:1-20.

[3] Simopoulos AP: The importance of the omega-6/omega-3 fatty acid ratio in cardiovascular disease and other chronic diseases. *Exp Biol Med* (Maywood) 2008, 233: 674-688.

[4] Venegas-Caleron M, Sayanova O, Napier JA: An alternative to fish oils: Metabolic engineering of oil-seed crops to produce omega-3 long chain polyunsaturated fatty acids. *Prog Lipid Res* 2010, 49: 108-119.

[5] Nykiforuk CL, Shewmaker C, Harry I, Yurchenko OP, Zhang M, Reed C et al.: High level accumulation of gamma linolenic acid (C18:3Delta6,9,12 cis) in transgenic safflower (*Carthamus tinctorius*) seeds. *Transgenic Res* 2011 DOI:10.1007/s11248-011-9543-5.

[6] Cheng B, Wu G, Vrinten P, Falk K, Bauer J, Qiu X: Towards the production of high levels of eicosapentaenoic acid in transgenic plants: the effects of different host species, genes and promoters. *Transgenic Res* 2010, 19: 221-229.

[7] Gatlin DM, Barrows FT, Brown P, Dabrowski K, Gaylord TG, Hardy RW et al.: Expanding the utilization of sustainable plant products in aquafeeds: a review. *Aquaculture Res* 2007, 38: 551-579.

[8] Plourde M, Cunnane SC: Extremely limited synthesis of long chain polyunsaturates in adults: implications for their dietary essentiality and use as supplements. *Appl Physiol Nutr Metab* 2007, 32: 619-634.

[9] Griffiths G, Morse N: Clinical applications of C-18 and C-20 chain length polyunsaturated fatty acids and their biotechnological production in plants. *J Am Oil Chem Soc* 2006, 73: 171-185.
[10] Uttaro AD: Biosynthesis of polyunsaturated fatty acids in lower eukaryotes. *IUBMB Life* 2006, 58: 563-571.
[11] Metz JG, Roessler P, Facciotti D, Levering C, Dittrich F, Lassner M et al.: Production of polyunsaturated fatty acids by polyketide synthases in both prokaryotes and eukaryotes. *Science* 2001, 293: 290-293.
[12] Griffiths G, Stobart AK, Stymne S: Delta 6- and delta 12-desaturase activities and phosphatidic acid formation in microsomal preparations from the developing cotyledons of common borage (*Borago officinalis*). *Biochem J* 1988, 252: 641-647.
[13] Tocher DR: Fatty acid requirements in ontogeny of marine and freshwater fish. *Aquaculture Res* 2010, 41: 717-732.
[14] Hong H, Datla N, Reed DW, Covello PS, MacKenzie SL, Qiu X: High-level production of gamma-linolenic acid in *Brassica juncea* using a delta6 desaturase from *Pythium irregulare*. *Plant Physiol* 2002, 129: 354-362.
[15] Liu JW, DeMichele S, Bergana M, Bobik E, Hastilow C, Chuang LT et al.: Characterization of oil exhibiting high gamma-linolenic acid from a genetically transformed canola strain. *J Am Oil Chem Soc* 2001, 78: 489-493.
[16] Eckert H, La VB, Schweiger BJ, Kinney AJ, Cahoon EB, Clemente T: Co-expression of the borage Delta 6 desaturase and the Arabidopsis Delta 15 desaturase results in high accumulation of stearidonic acid in the seeds of transgenic soybean. *Planta* 2006, 224: 1050-1057.
[17] Ruiz-Lopez N, Haslam RP, Venegas-Caleron M, Larson TR, Graham IA, Napier JA et al.: The synthesis and accumulation of stearidonic acid in transgenic plants: a novel source of 'heart-healthy' omega-3 fatty acids. *Plant Biotechnol J* 2009, 7: 704-716.
[18] Bell JG, Strachan F, Good JE, Tocher DR: Effect of dietary echium oil on growth, fatty acid composition and metabolism, gill prostaglandin production and macrophage activity in Atlantic cod (*Gadus morhua* L.). *Aquaculture Res* 2006, 37: 606-617.
[19] Abbadi A, Domergue F, Bauer J, Napier JA, Welti R, Zahringer U et al.: Biosynthesis of very-long-chain polyunsaturated fatty acids in transgenic oilseeds: constraints on their accumulation. *Plant Cell* 2004, 16: 2734-2748.
[20] Domergue F, Abbadi A, Ott C, Zank TK, Zahringer U, Heinz E: Acyl carriers used as substrates by the desaturases and elongases involved in very long-chain polyunsaturated fatty acids biosynthesis reconstituted in yeast. *J Biol Chem* 2003, 278: 35115-35126.
[21] Qi B, Fraser T, Mugford S, Dobson G, Sayanova O, Butler J et al.: Production of very long chain polyunsaturated omega-3 and omega-6 fatty acids in plants. *Nat Biotechnol* 2004, 22: 739-745.
[22] Wu G, Truksa M, Datla N, Vrinten P, Bauer J, Zank T et al.: Stepwise engineering to produce high yields of very long-chain polyunsaturated fatty acids in plants. *Nat Biotechnol* 2005, 23: 1013-1017.
[23] Kajikawa M, Matsui K, Ochiai M, Tanaka Y, Kita Y, Ishimoto M et al.: Production of arachidonic and eicosapentaenoic acids in plants using bryophyte fatty acid Delta6-desaturase, Delta6-elongase, and Delta5-desaturase genes. *Biosci Biotechnol Biochem* 2008, 72: 435-444.

[24] Hoffmann M, Wagner M, Abbadi A, Fulda M, Feussner I: Metabolic engineering of omega-3-very long chain polyunsaturated fatty acid production by an exclusively acyl-CoA-dependent pathway. *J Biol Chem* 2008, 283: 22352-22362.
[25] Dahlqvist A, Stahl U, Lenman M, Banas A, Lee M, Sandager L et al.: Phospholipid:diacylglycerol acyltransferase: an enzyme that catalyzes the acyl-CoA-independent formation of triacylglycerol in yeast and plants. *Proc Natl Acad Sci USA* 2000, 97: 6487-6492.
[26] Lu CF, Xin ZG, Ren ZH, Miquel M, Browse J: An enzyme regulating triacylglycerol composition is encoded by the ROD1 gene of Arabidopsis. *Proc Natl Acad Sci USA* 2009, 106: 18837-18842.
[27] Bates PD, Durrett TP, Ohlrogge JB, Pollard M: Analysis of acyl fluxes through multiple pathways of triacylglycerol synthesis in developing soybean embryos. *Plant Physiol* 2009, 150: 55-72.
[28] Bates PD, Browse J: The pathway of triacylglycerol synthesis through phosphatidylcholine in Arabidopsis produces a bottleneck for the accumulation of unusual fatty acids in transgenic seeds. *Plant J* 2011, 68: 387-399.
[29] Wood CC, Petrie JR, Shrestha P, Mansour MP, Nichols PD, Green AG et al.: A leaf-based assay using interchangeable design principles to rapidly assemble multistep recombinant pathways. *Plant Biotechnol J* 2009, 7: 914-924.
[30] Petrie JR, Shrestha P, Liu Q, Mansour MP, Wood CC, Zhou XR et al.: Rapid expression of transgenes driven by seed-specific constructs in leaf tissue: DHA production. *Plant Methods* 2010, 6: 8.
[31] Barker B. Brain-, eye-, and heart-healthy canola oil in the works. Top Crop Manager (http://www.topcropmanager.com/content/view/4128/38/) 2008. 5-15-2011.
[32] Nuseed. Australian scientific collaboration set to break world's reliance on fish for long chain omega-3. Nuseed media release. http://www.nuseed.com/Assets/448/1/ 110412_ Omega3inCanola_PioneeringAustralianresearchallianceannouncement_Embargo-Tuesday12April2011.pdf. 2011. 5-15-2011.
[33] Watkins C. Oilseeds of the future: Part 1. INFORM , May 2009, pp. 276-279.
[34] Monsanto. World's first SDA omega-3 soybean oil achieves major mileston that advances the development of foods with enhanced nutritional benefits. http://monsanto.mediaroom.com/index.php?s=43anditem=761andprintable 2009. 5-15-2011.
[35] Glencross B, Turchini G: Fish oil replacement in starter, grow-out, and finishing feeds for farmed aquatic animals. In *Fish Oil Replacement and Alternative Lipid Sources in Aquaculture Feeds.* CRC Press; 2010:373-404.
[36] Turchini GM, Torstensen BE, Ng WK: Fish oil replacement in finfish nutrition. *Rev Aquaculture* 2009, 1: 10-57.
[37] Das UN: Essential fatty acids - a review. *Curr Pharm Biotechnol* 2006, 7: 467-482.
[38] Mugrditchian DS, Hardy RW, Iwaoka WT: Linseed oil and animal fat as alternative lipid sources in dry diets for Chinook Salmon (*Oncorhynchus tshawytscha*). *Aquaculture* 1981, 25: 161-172.
[39] Stubhaug I, Lie O, Torstensen BE: Fatty acid productive value and beta-oxidation capacity in Atlantic salmon (*Salmo salar* L.) fed on different lipid sources along the whole growth period. *Aquaculture Nutr* 2007, 13: 145-155.

[40] Menoyo D, Lopez-Bote CJ, Diez A, Obach A, Bautista JM: Impact of n-3 fatty acid chain length and n-3/n-6 ratio in Atlantic salmon (*Salmo salar*) diets. *Aquaculture* 2007, 267: 248-259.

[41] Greene DHS, Selivonchick DP: Effects of dietary vegetable, animal and marine lipids on muscle lipid and hematology of rainbow-trout (*Oncorhynchus mykiss*). *Aquaculture* 1990, 89: 165-182.

[42] Yamazaki K, Fujikawa M, Hamazaki T, Yano S, Shono T: Comparison of the conversion rates of alpha-linolenic acid (18-3(N-3)) and stearidonic Acid (18-4(N-3)) to longer polyunsaturated fatty acids in rats. *Biochimica et Biophysica Acta* 1992, 1123: 18-26.

[43] James MJ, Ursin VM, Cleland LG: Metabolism of stearidonic acid in human subjects: comparison with the metabolism of other n-3 fatty acids. *Am J Clin Nutr* 2003, 77: 1140-1145.

[44] Harris WS, Lemke SL, Hansen SN, Goldstein DA, DiRienzo MA, Su H et al.: Stearidonic acid-enriched soybean oil increased the omega-3 index, an emerging cardiovascular risk marker. *Lipids* 2008, 43: 805-811.

[45] Brenna JT, Salem N, Sinclair AJ, Cunnane SC: alpha-Linolenic acid supplementation and conversion to n-3 long-chain polyunsaturated fatty acids in humans. *Prostag Leukotr Ess* 2009, 80: 85-91.

[46] Tocher DR, Dick JR, MacGlaughlin P, Bell JG: Effect of diets enriched in Delta 6 desaturated fatty acids (18:3n-6 and 18:4n-3), on growth, fatty acid composition and highly unsaturated fatty acid synthesis in two populations of Arctic charr (*Salvelinus alpinus* L.). *Comp Biochem Phys B* 2006, 144: 245-253.

[47] Miller MR, Nichols PD, Carter CG: Replacement of dietary fish oil for Atlantic salmon parr (*Salmo salar* L.) with a stearidonic acid containing oil has no effect on omega-3 long-chain polyunsaturated fatty acid concentrations. *Comp Biochem Phys B* 2007, 146: 197-206.

[48] Tocher D, Francis D, Coupland K: n-3 Polyunsaturated fatty acid-rich vegetable oils and blends. In *Fish Oil Replacement and Alternative Lipid Sources in Aquaculture Feeds*. CRC Press; 2010:209-244.

[49] Miller MR, Bridle AR, Nichols PD, Carter CG: Increased elongase and desaturase gene expression with stearidonic acid enriched diet does not enhance long-chain (n-3) content of seawater Atlantic Salmon (*Salmo salar* L.). *J Nutr* 2008, 138: 2179-2185.

[50] Alhazzaa R, Bridle AR, Nichols PD, Carter CG: Replacing dietary fish oil with Echium oil enriched barramundi with C-18 PUFA rather than long-chain PUFA. *Aquaculture* 2011, 312: 162-171.

[51] de Silva S, Soto D: Climate change and aquaculture: potential impacts, adaptation and mitigation. *FAO Fisheries Technical Paper* 2009, 530: 137-215.

[52] Mueller K, Eisner P, Yoshie-Stark Y, Nakada R, Kirchhoff E: Functional properties and chemical composition of fractionated brown and yellow linseed meal (*Linum usitatissimum* L.). *Journal of Food Engineering* 2010, 98: 453-460.

[53] Nennerfelt L, Evans D, Ten Haaf W.. Process for extracting flax protein concentrate from flax meal. Nutrex Wellness, Inc. [US 6998466 B2]. 2006. Patent

[54] Brunner G: Supercritical fluids: technology and application to food processing. *Journal of Food Engineering* 2005, 67: 21-33.

[55] Food and Agriculture Organization. The status of world fisheries and aquaculture. 2009. Rome, Food and Agriculture Organization. Tacon AGJ, Metian M: Global overview on the use of fish meal and fish oil in industrially compounded aquafeeds: Trends and future prospects. *Aquaculture* 2008, 285: 146-158.

In: Organic Chemistry, Biochemistry, Biotechnology ... ISBN: 978-1-62081-155-9
Editors: G. Zaikov, F. Pudel and G. Spychalski © 2013 Nova Science Publishers, Inc.

Chapter III

Potential and Future Prospects of Dyeing Plants: Madder *Rubia Tinctorium* L., Annato *Bixa Orellana* L., Dyer's Broom *Genista Tinctoria* L., Woad *Isatis Tinctoria*. L.

***Katarzyna Schmidt-Przewozna*[1], *Jakub Kowalinski*[1] and *Monia Tomassini*[2]**
[1]Institute of Natural Fibres and Medicinal Plants,
Wojska Polskiego Poznan, Poland
[2]University Camerino, Italy

Abstract

The material for obtaining natural dyes comes from fresh or dried plants traditionally known as dyeing plants that contain a pigment. A special group of these plants are those used for the production of food colourings. International standards allow only a few synthetic food colourings, preferring those of natural origin. This preference is linked with proven toxicity of synthetic dyes. Madder *Rubia tinctorium* L., Annato *Bixa orellana* L, Dyer's broom *Genista tinctoria* L., Woad *Isatis tintoria*. L are plant raw materials used for dyeing textiles and leather products, also used as food colourings, for the production of graphic, oil and water-colour paints, for painting on china and dyeing plaster.

Keywords: Madder, annatto, dyer's broom, woad, dyes, medicinal plants

Introduction

The study is an attempt to provide an eco-friendly substitute for synthetic dyes, the replacement from unconventional but natural sources. Colour obtained from nature has been part of our life from centuries.

Madder – the source of many shades of red colour, annatto – orange, yellow; dyers' broom – yellow, green, brown; Woad – blue are very important sources of colour. These plants, apart from dyeing properties, also have other features that determine their wide potential of use. They might be applied in cosmetic and pharmaceutical industries. Some of them have anti-inflammatory and antibacterial properties, which properties can be utilized in textile industry for the production of functional healthy clothing.

The paper will present the potential and future prospects of using dyeing plants.

Experimental

Departament of Natural Dyeing "Natural Art" conducts the program of dyeing by natural dyes. Every year in the experimental plantation in Petkowo we try to test new sources of plants with dyeing properties. A series of laboratory and pilot-scale extractions has been carried out including small agro-technical experiments, laboratory and pilot–scale extraction, dyeing tests and quality control. New dye crops offer farmers opportunities for diversification and can provide industry with new raw materials.

In our research program we also tested natural sources from different continents. The main idea of this presentation is to find different properties of the following dyeing plants: Madder *Rubia tinctorium* L., Annato *Bixa orellana* L, Dyer's broom *Genista tinctoria* L., Woad *Isatis tintoria*. L

1. Madder *Rubia Tinctorium* L.

1.1. Plant Characteristics and History

Madder *Rubia tinctorium* L. is one of the most ancient dyes. The genus Rubia belongs to the family Rubiaceae. It was known as early as 3.000 B.C. as a source of red colour. Madder has been cultivated as a source of dyestuff since antiquity in Europe, Middle East, Turkey, Central Asia and Egypt. [7]

Cloths dyed with madder root pigment was found in the tomb of the Pharaoh Tutankhamun, in the ruins of Pompeii, and ancient Corinth. In the Middle Ages, Charlemagne encouraged madder cultivation (Ball, 2002). In 1826, the French chemist Pierre-Jean Robiquet found that there were two colorants in madder root, *i.e.*, the red alizarin and a more rapidly fading purpurin. The alizarin as a synthetic dye was produced in 1868 by the German chemists Carl Graebe and Carl Liebermann, working for BASF.

The roots are harvested in the autumn from plants that are at least 3 years old.

The Rubia species have high commercial, economical, and medicinal importance. Plants belonging to this species are known to contain substantial amounts of anthraquinones, especially in the roots. Economically, the Rubia species are important as the source of potential biocolorants (natural dyes). They had long been used in India and elsewhere for imparting various shades of red, scarlet, light pink, salmon, coffee-brown and mauve to cotton, linen and woolen fabrics and others.

Figure 1. Madder *Rubia tinctorium* L. – experimental plantation INF&MP, Petkowo, 2010.

1.2. Madder *Rubia Tinctorium* L. - Natural Dye

The roots of this plant contain pigment alizarin and purpurin, lucidin, pseudopurpurin, rubiadin, and the glucosides.

1.3. Method of Dyeing

Extraction of Dye

The plants are crushed to small pieces and soaked in hot water overnight, boiled one hour and filtrated.

Mordants

In our methods we used: oak galls, bark of oak, Sodium carbonate anhydrous, Copper sulphate, Citric acid, copper sulphate, Iron - ferrous sulphate, Alum-potassium aluminium sulphate

Development of Color on Linen, Hemp and Silk Fabrics

Equipment: Laboratory dyeing machine: EASYKROME UGOLINI

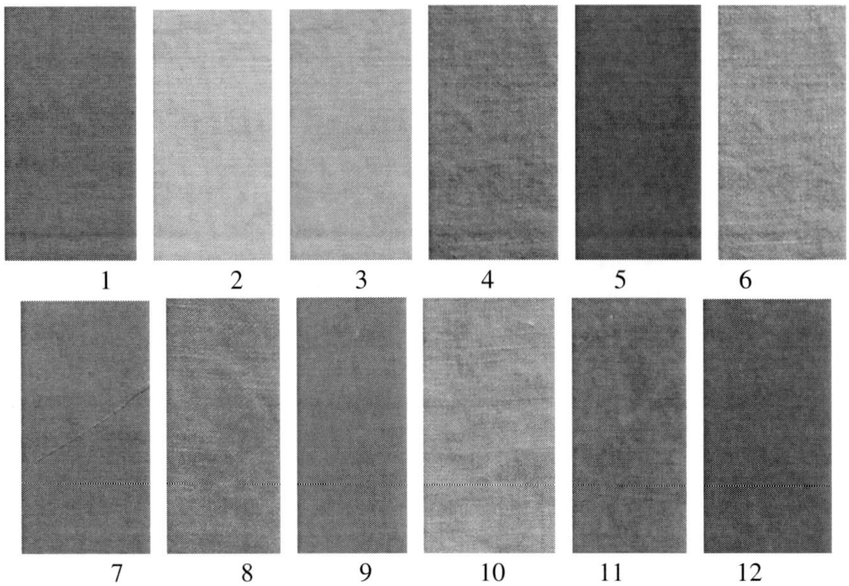

Figure 2. Colour result – Madder dyeing.

Table 1. Madder dyeing results on linen, silk samples

No.	Fabric	mordant	colour	Wash fastness fastness	Light fastness
1	Silk B	No mordant	orange	4	4
2	Linen A	No mordant	light orange	4	4
3	Linen A	0,3% alum	light orange	5	6
4	Linen C	0,3 % soda	light violet	5	6
5	Silk D	0,3 % soda	dark maroon	5	6
6	Linen A	0,3 % citric acid	light pink	4	4
7	Silk F	0,3 % citric acid	redish orange	4	5
8	Linen C	0,3 % copper sulphate	rosybrown	5	5
9	Silk D	0,3 % copper sulphate	dark rosybrown	4	5
10	Linen E	0,3 % ferrous sulphate	rosybrown	5	5
11	Linen C	0,3 % ferrous sulphate	rosygray	5	5
12	Silk D	0,3 % ferrous sulphate	dark rosybrown	5	5

A - linen no mordant
B - silk no mordant
C - linen premordating in galas
D - silk premordating in galas
E - linen premordating in bark of oak
F - silk premordating in bark of oak

Extraction of Dyestuff from Madder Root

Soxhlet apparatus was used to extract the dyestuff from *Rubia tinctorum* L., Dried and grounded roots were extracted with ethanol. After extraction the solvent was evaporated with rotavapor and a very viscous, dark and sweet smelled extract was obtained. For dry the extract another apparatus (spray dryer) was used and an orange powder was obtained. This powder consist on dyes and other cellular constituent and can be used for dyeing.

All the extract have been evaporated with:

- Vacum Evaporator R-210
- BUCHI Mini Spray Dryer B-290

Figure 3. Extraction of dyestuff from Madder root.

Reasult

Extractions Result

Soxhlet

Powder from dryer for each sample:

1) 1,9985 g (from 31,77 g of madder) = 6,3103%
2) 2,0991 g (from 45,03 g of madder) = 4,6615%
3) 3,0604 g (from 35,2036 g of madder) = 8,6934%
4) 2,9946 g (from 37,8177 g of madder) = 7,9185%

With the dryer is impossible recover the total amount of extract, a part of this still remain on the wall of the machine.

Medicinal Use

The medicinal part of *Rubia tinctorum* is the dried root (Blumenthal *et al.*, 1998).

The roots of madder were used in the Ayurvedic and Unani system of traditional medicine In India and in other parts they were used for dissolving kidney stones, against paralysis, urinary disorders, herpes simplex infection, skin conditions (Brown &Brown, 1976; Mori *et al.*, 1990; Wang *et al.*, 1992). Madder is also endowed with genotoxic activity. The madder extract have the antimicrobial and anti-fungal and anti-inflammatory properties.[2,6]

In South Africa it is used as aphrodisiac and to treat impotence. In Ethiopia the roots are used to treat amoebic dysentery, cancer and coughs. [3]

Alizarin is not recommended for pregnant and breast feeding women.

Food Colorant

The extract of *Rubia tinctorum* L has been used for colouring chewing gums, noodles and salted foods.

Madder colour (MC) extracted from the root of Rubia tinctorum L. was long accepted for use as a food additive in Japan (MHLW,1995) and Korea, and has been widely applied for a variety of foods/drinks, such as confectionaries, processed meats (hams and sausages), boiled fish paste and beverages. Major constituents of MC are anthraquinones, such as lucidin, alizarin and their primeverosides, e.g., lucidin-3-O-primeveroside and ruberythric acid [2]. Ministry of Health, Labour and Welfare in Japan supported Grant-in-Aid for the study titled: Induction of kidney and liver cancers by the natural food additive madder colour in a two-year rat carcinogenicity study. The research was carried out on groups of rats (344 rats in the two-year-experiment). Histopathological changes suggestive of renal carcinogenicity were found only in 5,0 % tested animal but survival in those group was higher than in others.[4]

Cosmetics

Extract obtained from roots of Madder is used in shampoos, lotions and soap**.** This extract has antibacterial properties.[6]

2. Annato *Bixa Orellana* L.

2.1. Plant Charatteristic and History

The plant grows in the tropical area. Annatto is found in Ecuador, Mexico, Bolivia and Brazil. Annatto's Latin designation (*Bixa orellana* L.) was named after the Spanish conquistador Francisco de Orellana during his exploration of the Amazon River.[8]

Annatto is a profusely fruiting shrub or a small tree that grows 5-10 m in height. Approximately 50 seeds grow inside prickly reddish-orange heart-shaped pods. One small annatto tree can produce up to 270 kg of seeds.

Pigment from seeds of this tree contains bixine and norbixine, natural colorants used for food industry and as natural dyes for textiles and also for coloring feathers, sheepskin, ivory, bones, bamboo mats and rattan. The color may be fixed with tamarind leaves or with barks containing tannin. [9] These pigments are carotenoids that vary in tone between yellow and red colours. Specialist literature reports that from leaves we can obtain green shades.

2.2. Annato *Bixa Orellana* L. - Natural Dye

In addition to bixin and norbixin, annatto contains bixaghanene, bixein, bixol, crocetin, ellagic acid, ishwarane, isobixin, phenylalanine, salicylic acid, threonine, tomentosic acid, and tryptophan.

Figure 4. Annato Bixa orellana L.- seeds.

2.3. Method of Dyeing

Extraction of Dye

The powder from seeds and soaked in hot water overnight, boiled one hour and filtrated.

Mordants

In our methods we used: oak galls, bark of oak, sodium carbonate anhydrous, copper sulphate, citric acid, copper sulphate, Iron - ferrous sulphate, Alum-potassium aluminium sulphate

Development of Color on Linen, Hemp and Silk Fabrics

Equipment: Laboratory dyeing machine: EASYKROME UGOLINI

Parts Used: Seeds

Medicinal

Non-carcinogenic and non-toxic native plant, used as haemostatic, *antioxidant*, astringent *and antibacterial,* antidysenteric, diuretic, aphrodisiac and an effective febrifuge, digestive and gentle purge. It is prescribed also for epilepsy, erysipelas and sundry skin diseases, throat infections.

Seeds contain characteristic pleasant-smelling oil. The composition of the seed oil is similar to that of soybean oil. Fats in plant can be used for production of butter. [9] A decoction of Annatto root is taken orally to control asthma. A macerated seed decoction is taken orally for relief of fever, and the pulp surrounding the seed is made into an astringent drink used to treat dysentery and kidney infection. Infusions of root in water and rum are used to treat venereal diseases. The dye is used as an antidote for prussic acid poisoning. Extracts from different parts of the tree kill bacteria, fight free radicals, kill parasites, increase urination, stimulate digestion, lower blood pressure, act as mildly laxative, protect liver, reduce fever, arterial hypertension, high cholesterol, cystitis, obesity, renal insufficiency, and eliminate uric acid.

Table 2. Annato dyeing results on linen, silk samples

No.	Fabric	mordant	Colour	Wash fastness	Light fastness
1	B	No mordant	orange	4	4
2	A	No mordant	Light orange	4	3
3	A	0,3% alum	orange	5	5
4	D	0,3% alum	Bright orange	5	5
5	C	0,3 % soda	Dark orange	5	5
6	A	0,3 % soda	Light coral	5	4
7	F	0,3 % citric acid	Light peach	4	4
8	C	0,3 % copper sulphate	peach	5	5
9	A	0,3 % copper sulphate	Light brown	4	5
10	E	0,3 % ferrous sulphate	Light brown orange	5	5
11	C	0,3 % ferrous sulphate	Redish brown orange	5	5
12	F	0,3 % ferrous sulphate	brown orange	5	5

A - linen no mordant
B - silk no mordant
C - linen premordating in galas
D - silk premordating in galas
E - linen premordating in bark of oak
F - silk premordating in bark of oak

Food Colorant

Cheese, sausages, meats and candies and also beverages, cosmetics, pharmaceutical products, traditional spices for food.

Cosmetics

Annatto has a common name "lipstick tree" because in the past it was in use for colouring women's lips and for body painting. From seeds we can obtain dye and oil extract for cosmetics, essential oils for shampoos, soaps etc. From colour cosmetics list code of annatto seeds is the same like extract from annatto. Code is: E160b; CI number is 75120.

In Department of Natural Dyeing in INF&MP we obtain reddish orange to yellow hues on silk, linen, cotton, hemp and wool samples. For tested samples we obtain very good results for wash and light fastness. Reddish orange and yellow on our dyed materials proved that annatto is very stable colorant.

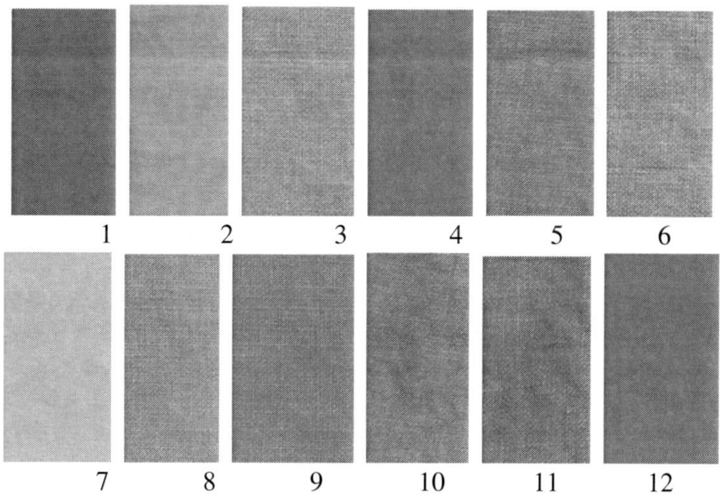

Figure 5. Colour result – Annato dyeing.

3. Dyer's Broom *Genista Tinctoria* L.

3.1. Plant Charatteristic and History

Dyer's Broom (Dyer's Greenweed, Dyer's Greenwood, Woadwaxen)

The genus belongs to family family Fabaceae. The plant is commonly found on dry soil and meadows in to the Mediterranean region, especially in Europe, and the some regions of Asia and North America. [7]

Dyer's broom is small Eurasian shrub having clusters of yellow flowers that yield a dye. The traces of *Genista tinctoria* L. have been found on archeological sites. The name of Plantagenet kings family is coming from the dyer's broom known as as *"planta genista"* in Latin. The Geffrey of Anju (father of Henry II of England) took the plant as an emblem of the family.

Historically it was one of the main sources of yellow colour and with blue (from woad *Isatis tinctoria* L.,) created green. – it gives the name to the plant "woadwaxen". In the past the Romans had extensively used the dyer's greenweed to dye clothes and leather in the same procedure and this has been documents by several ancient Roman writers. Combinations of yellow coming from dyer's broom with blue obtain from woad *Isatis tinctoria.* L produced famous colour in history Kendal blue. This colour was popular in history for forest man and army costumes.

"How couldst thou know these men in Kendal green, when it was so dark thou couldst not see thy hand?"—Shakespeare: 1 Henry IV., ii. 4.

The main component of the plant are: flavone, luteolin, apigenin, genistin.[12]

3.2. Dyer's Broom - Natural Dye

Dye Composition

The main colorant is: flavone, luteolin.

Table 3. Dyer's broom dyeing results – Genista tinctoria L.

No.	Fabric	mordant	Colour	Wash fastness	Light fastness
1	B	0,3% alum	Light green yellow	4	5
2	D	0,3% alum	Olive green yellow	5	5
3	E	0,3% alum	Green yellow	5	5
4	A	0,3% alum	Green yellow	5	6
5	C	0,3% soda	Light khaki	5	4
6	D	0,3 % soda	Khaki	5	5
7	D	0,3 % copper sulphate	Yellow olive	5	5
8	C	0,3 % copper sulphate	Brown	5	5
9	B	0,3 % copper sulphate	Light rosybrown	4	5
10	D	0,3 % ferrous sulphate	rosybrown	5	5
11	F	0,3 % ferrous sulphate	Dark rosybrown	5	5
12	C	0,3 % ferrous sulphate	Dark brown	5	5

A - linen no mordant
B - silk no mordant
C - linen premordating in galas
D - silk premordating in galas
E - linen premordating in bark of oak
F - silk premordating in bark of oak

All parts of the plant, but especially the flowering tops, yield a good yellow dye, and from the earliest times have been used by dyers for producing this colour, especially for wool;

In history for dyeing experiment all parts of plant were used. Dyers'bromm is an source of yellow colour. The plant tops were responsible for bright yellow. With addition copper sulphate is possible to obtain green. In UK Kendal green is a local common name for the plant. The brighter shade of “Kendal” is "Saxon green". In surrounding of Leipzig the Saxon blue and Saxon green cloths were popular he in XVII century.[11]

Medicinal

The medicinal parts are the flowering twigs, seeds and the ashes form an alkaline salt. *Genista tinctoria* has diuretic, cathartic, aperient and emetic properties.

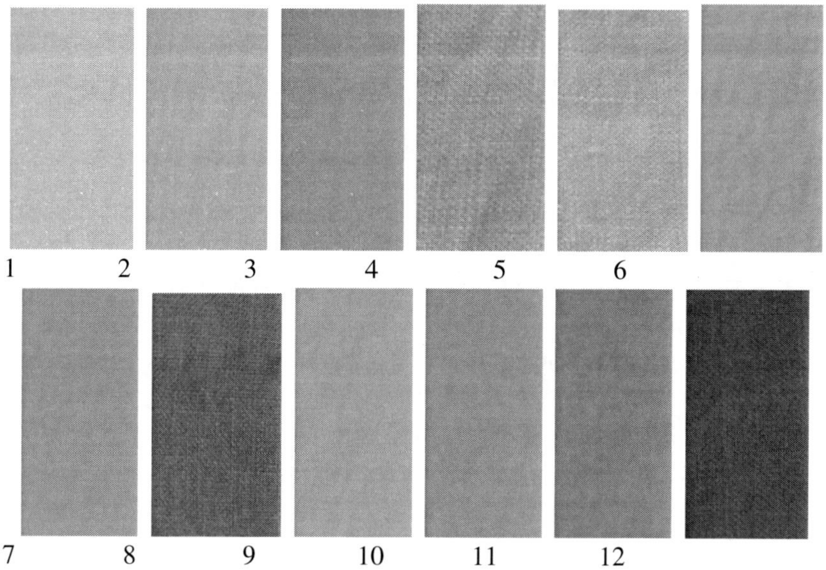

Figure 6. Colour result – Dyer’s broom dyeing.

Genista tinctoria L. contains important quantities of phytoestrogens such as: genistein, daidzein, daidzin formononetin, ononin. Most of the pharmacologic studies use as source of phytoestrogens, soy (Glycine max) or red clover (Trifolium pratense), but an alternative source can be Genista tinctoria L Isoflavones are natural agonists or antagonists of estrogenic receptors E_2. Moreover, they inhibit protein tyrosine kinases which participate to the growth factor stimulated transduction cascades in normal and transformed cancer cells. That is why isoflavones are believed to inhibit the formation of tumors, especially those which are related to the hormonal balance of the body (breast cancer, ovaries cancer, prostate cancer).

Isoflavonoids also diminish the symptoms of menopause such as osteoporosis and cardiovascular diseases [1, 6]. In fusion is used against rheumatism and gout dropsy and chronic skin disordes. The extract is used for herpes. It is not good for the peoples with high blood preasure.[12]

Cosmetics and Food Products

Genistin have positive influence for skin collagen which is good for anti-ageing products.

In adipose tissue it produces a lipolytic activity is so useful in the treatment of cellulite. It is component of creams and lotions. [6]

In Colour list genistin coming from Dyer’s broom giving yellow colour have number 75610.

4. Woad *Isatis Tinctoria* L.

4.1. Plant Characteristics and History

4.2. Woad Isatis Tinctoria L. is a Flowering Plant belonging to the Family Brassicaceae

Specialist literature reports that woad has been known as a dye plant since the Neolithic age. Ancient Britons and Celts used woad to colour their faces and bodies with the extract from leaves in order to frighten their enemies. In the Middle Ages the trade of this precious blue dye became the basis of the European economy.

Its importance began to decline during the 16th century, when a cheaper blue dye derived from *Indigofera tinctoria* L. was imported from the Far East. By the beginning of 20th century, the cultivation of woad had almost disappeared in Europe [7,14].

Figure 7. Woad Isatis tinctoria L.- experimental farm INF&MP, Petkowo.

4.3. Compounds

Glucobrassicin, Indoxyl-Beta-Glucoside, Beta-Sitosterol, Isatin, and Indirubin

4.4. Woad *Isatis tinctoria* L. – Natural Dye

There is a plot of woad cultivated at the Experimental Farm of INF&MP in Petkowo. It is sown in early spring and does not pose any problems in cultivation. The leaves that contain blue pigment are harvested in summmer, until the middle of autumn. The seeds that contain grey pigment are harvested in autumn.

The project titled: Ecologization of agro-tourist farms, Natural Art Studio conducted dyeing workshop that included producing several trials on flax, hemp and silk yarns and fabrics.

Figure 8. Woad Isatis tinctoria L.- experimental farm INF&MP, Petkowo.

A number of cotton t-shirts were produced in the tie-dye technique, which were then tested for light and washing fastness according to an international standard PN-EN ISO 105-C06:2010. The dyes examined exhibited very good wash and light fastness.

Table 4. Woad dyeing results – Isatis tinctoria L.

No.	Fabric	mordant	colour	Wash fastness	Light fastness
1	A	Vat dyeing	Midnight blue	5	5
2	C	Vat dyeing	Dark blue	5	5
3	E	Vat dyeing	Grey blue	5	4
4	B	Vat dyeing	Navy blue	5	5
5	D	Vat dyeing	Blue	5	5
6	E	Vat dyeing	Light blue	5	4

A - linen no mordant
B - silk no mordant
C - linen premordating in galas
D - silk premordating in galas
E - linen premordating in bark of oak
F - silk premordating in bark of oak

Figure 9. Linen abs silk samples dyed – woad.

Figure 10. Cotton and linen fabrics dyed woad.

Figure 11. Cotton and linen fabrics dyed woad.

4.5. Medicinal

Isatis leaves and roots have been used in traditional medicine mainly for treatment of infections; specifically, encephalitis, upper respiratory infection, and gastroenteritis, solid tumors and leukemia, the traditional usage that led to purification of the component compound indirubin. In Chinese medicine woad was used to clear heat and toxins from the blood [14]. Woad has 65 times more glucobrassicin than broccoli thus it can be used in cancer prevention. Woad is known for its antibacterial, antiviral, and antiparasitic activity.

4.6. Cosmetics

We can obtain oil extract from woad seed for cosmetics, essential oils for shampoons, soaps and hair dyes.

There is increasing interest in adding value to textiles by the use of natural products. Many of the plants used for obtaining natural dyes are additionally known to have medicinal properties.

The dyes examined exhibited good wash fastness and the antibacterial effect is therefore likely to be durable.

Antibacterial finishes are also highly desirable for textiles in the medical field as textile materials used in hospitals and hotels.

A number of studies are mentioned in the literature in which medicinal plants have been tested for activity against bacteria, fungi and viruses with a view to exploring their potential medicinal value [5,8]

References

[1] Chopra, Ananda S. (2003). *"Āyurveda"*. In Selin, Helaine. Medicine Across Cultures: *History and Practice of Medicine in Non-Western Cultures.* Norwell, MA: Kluwer Academic Publishers. pp. 75–83. ISBN 1-4020-1166-0.

[2] Blömeke, B., Poginsky, B., Schmutte, C., Marquardt, H. & Westendorf, J. (1992) Formation of genotoxic metabolites from anthraquinone glycosides present in *Rubia tinctorum L. Mutat. Res.,* 265, 263–272.

[3] Zimudzi C , Cardon D (2005) *Morinda luciada, in Japnsen and Cardon* 2005 pp110-12 2006.

[4] Inou K,, Yoshida M, (2009) "Induction of kidney and liver cancers by the natural food additive madder color in a two-year rat carcinogenicity study", *Food and Chemical Toxicology* 47, pp 184–191.

[5] D Gupta, Kumar S and Laha A, "Untimicrobial properties of natural dyes against Gram-negative bacteria" *Coloration Technology,* 2004 no 120 pp 166-171.

[6] Dweck A. C "Natural ingredients for colouring and styling" 2002, *International Journal of Cosmetic Science,* 24, pp. 287-302.

[7] Cardon D (2007) *Natural Dyes: Sources, Tradition, Technology and Science,* Archetype Publications, UK.

[8] Lauro, Gabriel J. and Francis, F. Jack. (2000). *Natural Food Colorants.* New York; Science and Technology: IFT Basic Symposium Series, Marcel Dekker, Inc.

[9] Julia F. Morton. (1960). *Can annatto (bixa orellana l), an old source of food color.* Meet new needs for safe dye, Florida State Horticultural Society, USA.

[10] G. F. Silva, Felix M. C. Gamarra, A. L. Oliveira and F. A. Cabral. (2008). *Extraction of bixin from annatto seeds using supercritical carbon dioxide.*

[11] R Forberger, *Die Manufaktur in Sachsen vom Ende des 16. bis zum Anfang des 19. Jahrhunderts* (Berlin, 1958), pp.149.

[12] Tero-Vescan A., Imre S.*,"Determination of some isoflavonoids and flavonoids from Genista tinctoria L by HPLC-UV"* Farmacia (2009),vol LVII, 1 pp 120-121.

[13] Santell R.C., Chang Y.C., Nair M.G. and Helferich W.G, Dietary Genistein Exerts Estrogenic Effects Upon the Uterus, Mammary Gland and the Hypothalamic/Pituitary axis in Rats. *J Nutr.,* 1997, 127(2), 263-269.

[14] Stefania Galletti, Jessica Barillari,Renato Lori and Gianpietro Venturi. (2006). Glucobrassicin enhancement in woad (Isatis tinctoria) leaves by chemical and physical treatments. *Journal of the Science of Food and Agriculture* ISBN:86:1833–1838 pp.1-7.

[15] Yeung HC. *Handbook of Chinese Herbs and Formulas*. Los Angeles, CA: No publisher listed; 1985. pp. 1.

[16] Cardon, D., *"Natural Dyes, Our Global Heritage of Colors"* (2010). Textile Society of America Symposium Proceedings.

In: Organic Chemistry, Biochemistry, Biotechnology ... ISBN: 978-1-62081-155-9
Editors: G. Zaikov, F. Pudel and G. Spychalski © 2013 Nova Science Publishers, Inc.

Chapter IV

Genetic Improvement of Oat Hulls and their Potential for Use as a Renewable Resource

Steffen Beuch[1], Bodo Saake[2], Jürgen Puls[3], Dirk Kahlke[4] and Stefan Geiser[4]

[1]Nordsaat Saatzucht GmbH, Breeding Station Granskevitz, Germany
[2]Department of Wood Science, Chemical Wood Technology, University of Hamburg, Germany
[3]von-Thünen-Institut (vTI), Institute for Wood Technology and Wood Biology Hamburg, Germany
[4]Peter Kölln KGaA, Köllnflockenwerke Elmshorn, Germany

Abstract

Oat hulls remain as a by-product after the milling of covered oat grains. In Germany commonly they are used as a fibre component in the feeding industry. To higher the economical output of the oat milling process and to meet the demands of more severe EU laws alternatives of this kind of usage are required. Oat hull xylan (representing about 30 % of dry matter of the hull) offers the potential for high value end use in different technical ways (i.e. pulp and paper, pharmaceuticals, extra food components). But especially a high lignin content of the hull disturbs the ecological and economical sustainability of the whole xylan extraction process.

Spring oat field trials with adapted milling oat varieties as well as low lignin/high xylan oat hull breeding lines have been established in Central Europe over three years between 2007 and 2009. They have shown the potential of a genetic increase of the sustainability of the whole process chain.

The lignin content of the oat hull was more stable over years and locations than the xylan content. In comparison with conventional spring oat varieties a reduction of the hull lignin content of more than 50 % is possible using special donors. It is followed by even a higher xylan content and a clear better processing ability. Four types of the inheritance of hull lignin and hull xylan clearly could be identified. But to fully compete with normal milling oat varieties a further increase of the grain yield and a better performance of some agronomic characteristics of these new oat breeding lines are essential. It can be assessed that a release of fully competitive low lignin/high xylan oat hull varieties is possible in the near future. The utilisation of the process residue (about 40 % of the oat hull) still remains open.

Introduction

Oats (*Avena sativa* L.) have been a major crop in European Agriculture for several years. About 1900 there were an oat cultivation of 4 Mio. hectares in the German Empire alone (1). The cultivation area of the oat crop have been decreased dramatically meanwhile but there were remarkable differences region by region and country by country. Today the best conditions for oat growing world wide can be found under autumn sowing in Western Europe (2). It is obvious that there is a clear international trend for using more oats for human consumption whereas the feed use of oats has been decreased rapidly the last years (3). Therefore breeding for a higher value of oats for food processing is essential. The nutritional ingredients of the oat grain have been in the major focus of oat breeders up to now because there is scientific evidence for oat food as healthy novel food according to "Health claim" laws (4,5,6). Up to now the oat breeding for a better technical processing ability mostly includes the look for a better outside grain quality (grain size, grading, hull content, dehullability, specific weight). A report of concrete practical oat breeding for non food purposes in the chemical industry has never been given before.

Aim and Background

There is an annual oat production of about 6 Mio. tons in the EU 27 presently. In Germany approximately 800.000 tons of oats are harvested year by year. The German oat mills belong to biggest oat processors all over Europe. They have an annual demand of about 300.000 tons of oats where about 80.000 tons of oat hulls remain as residues after the milling process. The German feed industry is the main processor of oat hulls using the hulls mainly as a fibre component. To meet the demands of more severe EU laws and to higher the economical output of oat processing alternatives of this kind of usage are required. The look for a higher value oat included the structural components of the oat hull where the arabinoxylan, a polysaccharide, seemed to be most promising for the use in the pulp and paper industry. Pre-investigations have shown the potential of the oat hull xylan as an additional component for the production of pulp and paper (7,8).

Therefore the aim of the research consortia was the development of technically better adapted oat varieties to higher the sustainability of the chemical hull processing. Especially a high lignin content of the oat hull disturbs the processing ability (7). In the past the hull lignin content of oats already has been of interest to oat breeders in order to increase the digestibility for better feed use. For example the Canadian spring oat variety AC Assiniboia has been described to have a clear lower lignin content of the hull (9). The inheritance of its hull lignin content is rather simple (mainly monogenetic) and (not very closely) coupled to phenotypic (brown hull colour) and genotypic (RFLP) markers (10). So the variety seemed to be a promising quality donor in breeding oats for this special technical purpose. Consequently oat breeding lines (mainly out of crossings with AC Assiniboia) with a possible lower lignin content of the oat hull have been in the major focus of the research work.

Material and Methods

Between 2007 and 2009 oat field trials have been performed at eight locations in Central Europe (6 Germany, 1 France, 1 Poland). The locations were part of the Nordsaat oat field trial network and represented various climatic conditions. A high adaptability of newly released oat varieties is essential for its market success (2).

Every year the variety set included three standard varieties, the donor variety AC Assiniboia, 11 milling oat varieties and 10 breeding lines (partly changed year by year). They have been tested in a 5x5-lattice design with two replications. The spring oat varieties represented about 20 % of the whole European spring oat area 2010. An overview of the trial program is given in table 1.

The field trials have been established and scored according to the field trial guidelines of the German National Plant Variety Board (11). For the statistical analysis the software package PLABSTAT have been used (12).

Regular samplings of the trial plots have been investigated for inside and outside grain quality parameters as well as the hull consistence. Before starting the chemical analysis the oat hulls have been sieved in order to be free of sugar and starch residues. For a more unique consistence all hull samples have been milled (UZM Retsch, Germany). The chemical oat hull structure has been analysed via a total hydrolysis with sulphuric acid followed by filtration and gravi-metric as well as UV-spectroscopic detection of the lignin. Additionally the content of different monomer carbohydrates was measured with the help of anionic exchange chromatography (13,14,15). For practical plant breeding purposes a validation of the NIRS-method to the xylan and lignin content of the oat hull have been proofed.

Results and Discussion

The testing period between 2007 and 2009 included various growing conditions at the different trial sites. Record yields up to 10 tons per hectare but also a very low potential (mainly caused by drought or heavy lodging) have been scored. However, the different vegetation caused a lot of valuable breeding information. The table 2 shows a summary of the agronomic performance of the different oat genotypes.

It is easy to see that the selected oat lines as well as the donor AC Assiniboia often showed a negative agronomic performance like lower grain yield, later panicle emergence, greater plant length, delayed straw ripening and stronger susceptibility to mildew (red and green coloured). Additionally for a correct assessment of its competitive lodging resistance their lower yield potential has to be taken into consideration (table 2).

Table 1. Overview field trial program

Year	Locations	Standard varieties	Milling varieties	Breeding lines
2007	- Granskevitz (DE) - Gudow (DE) - Böhnshausen (DE) - Moosburg (DE) - Güsten (DE) - Weimar-Schöndorf (DE) - Estrees St. Denis (FR) - Spytkowki (PL)	- Aragon - Flämingsprofi - Dominik - AC Assiniboia	Jumbo, Lutz, Nelson, Duffy, Ivory, Carron, Typhon, Pergamon, Scorpion, Husky, Buggy	07/1416 to 07/1425
2008	- Granskevitz (DE) - Gudow (DE) - Böhnshausen (DE) - Moosburg (DE) - Güsten (DE) - Weimar-Schöndorf (DE) - Estrees St. Denis (FR) - Spytkowki (PL)	- Aragon - Flämingsprofi - Dominik - AC Assiniboia	Jumbo, Lutz, Nelson, Duffy, Ivory, Carron, Typhon, Pergamon, Scorpion, Husky, Buggy	07/1417, 07/1418, 07/1420, 07/1422, 07/1425; 08/1421 to 08/1425
2009	- Granskevitz (DE) - Gudow (DE) - Böhnshausen (DE) - Moosburg (DE) - Güsten (DE) - Weimar-Schöndorf (DE) - Estrees St. Denis (FR) - Spytkowki (PL)	- Aragon - Flämingsprofi - Dominik - AC Assiniboia	Jumbo, Lutz, Nelson, Duffy, Ivory, Carron, Typhon, Pergamon, Scorpion, Husky, Buggy	07/1422, 07/1425; 09/1418 to 09/1425

At the other hand the variation of the height of the single traits offers the possibility to overcome these disadvantages during the following recurrent cycles of selection. So the selection progress was 3 % per year for the grain yield alone. This is more than the value of 0.8 % per year calculated for the grain yield breeding progress in the German National List trials with spring oats during the last 15 years (2).

The results of the grain quality investigations are summarised in table 3. In contrast to the agronomic value the grain quality of the breeding lines could compete with domestic milling oat varieties.

It is caused by the above average milling grain quality of the donor variety that is recommended as a milling oat variety by the oat mills in Canada. However, its kernel yield potential was still clearly lower in comparison with adapted milling oat varieties (red coloured) because it is determined by the grain yield alone if the hull content is on a comparable level. In spite of its low importance at a wide medium level for the milling ability of oats also the specific weight of the breeding lines is needed to be increased. This trait still has big importance for European oat trading. Considering the variability of the grain quality characteristics the same development like for the agronomic traits can be expected during the next recurrent breeding cycles.

Table 2. Grain yield and agronomic traits 2007-2009

	Yield (dt/ha)	Yield (%)	Panicle emergence (date)	Plant length (cm)	Lodging at harvest (1-9)	Straw ripening (1-9)	Suscep. to mildew (1-9)
	n=24	n=24	n=17	n=23	n=11	n=8	n=10
Standards (n=3)	72.6	100	10/06	107	3.2	4.3	3.7
AC Assiniboia	55.0	76	13/06	116	3.0	5.3	5.3
Breeding lines Av.	64.6	89	12/06	111	3.0	4.9	3.9
Breeding lines Min.	60.2	83	07/06	99	1.8	3.6	2.7
Breeding lines Max.	68.1	94	15/06	119	4.1	6.4	5.8
Varieties Av.	70.8	98	09/06	105	3.2	4.1	3.2
Varieties Min.	66.4	92	07/06	80	1.2	3.3	1.8
Varieties Max.	74.3	103	13/06	112	5.9	5.1	4.9

Table 3. Grain quality traits 2007-2009

	Kernel yield (dt/ha)	Kernel yield (%)	TGW (g)	Hectolitre weight (kg)	Hull content (%)	Kernel crude fat (%)	Kernel crude protein (%)	Kernel ß-Glucan (%)
	n=23	n=23	n=23	n=23	n=23	n=23	n=23	n=23
Standards (n=3)	55.2	100	37.7	49.6	25.0	6.8	15.6	4.1
AC Assiniboia	41.9	76	39.7	47.2	23.6	7.6	16.5	4.3
Breeding lines Av.	49.1	89	36.6	48.2	24.6	6.6	15.7	4.0
Breeding lines Min.	44.4	80	32.3	46.0	23.0	5.8	14.4	3.5
Breeding lines Max.	53.1	96	42.5	50.9	26.1	8.2	17.6	4.6
Varieties Av.	54.6	99	37.8	50.4	24.1	6.4	15.6	4.1
Varieties Min.	51.3	93	32.1	46.4	22.6	5.4	14.5	3.7
Varieties Max.	57.3	104	43.9	53.3	25.9	8.6	16.2	4.6

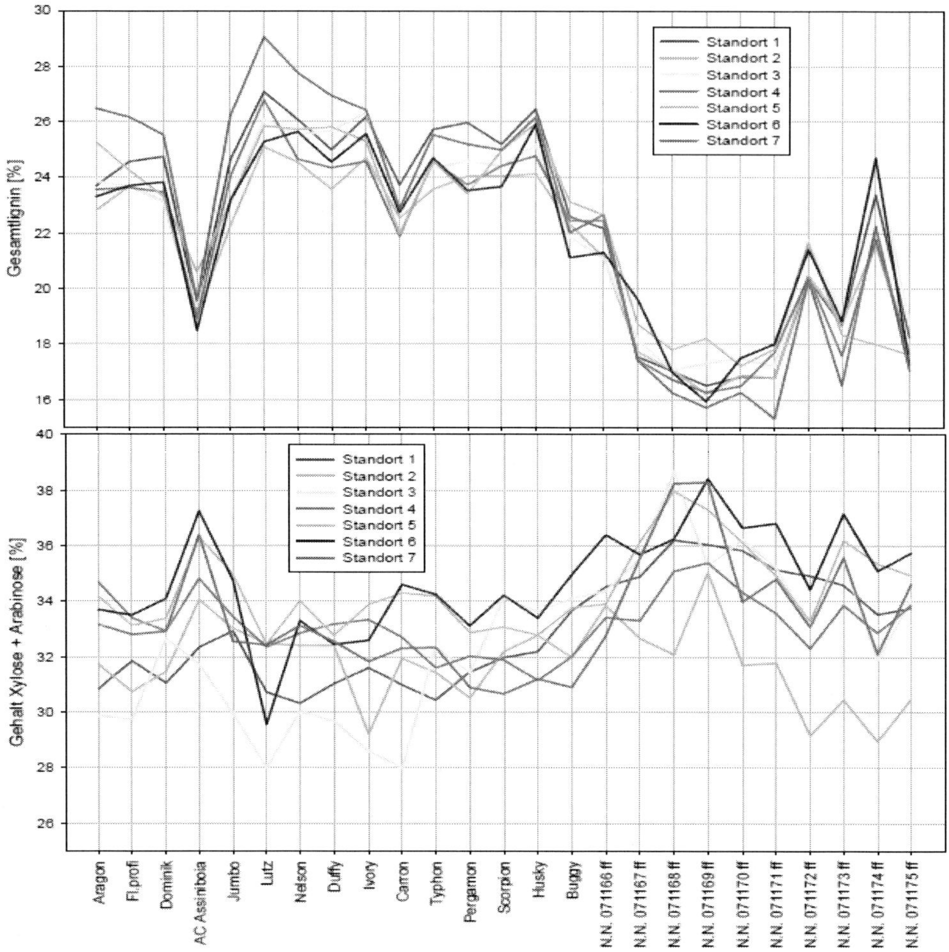

Figure 1. Hull lignin (above) and arabinoxylan (below) content 2007.

The figures 1 and 2 show a comparison of the hull lignin and the hull arabinoxylan content in the both years 2007 and 2008. According to the results of the hull analysis many of the new breeding lines had the expected consistence with lower lignin and higher xylan contents. In a few cases even a transgressive increase in comparison with the donor variety was obvious. The height of both lignin and arabinoxylan content has been influenced by the year and the location too. In 2007 the best lines had a hull lignin content of about 16 % whereas in 2008 some lines reached 15 % or even less. These low values also have been detected in 2009. The highest hull arabinoxylan content reached about 38 % in 2007 but about 45 % in 2008. In 2009 the hull arabinoxylan level was again even to 2007. It is assumed that the special growing conditions (especially drought combined with lower yields and poor outside grain quality) were responsible for a lower lignin and a higher xylan content of the oat hull in 2008. Over all three years the hull lignin content was more stable to the impact of environmental conditions (location, year) than the hull arabinoxylan content.

According to the genetic impact to the hull lignin and the hull sugar content four types clearly could be separated. Conventional oat varieties (type 1) varied between 23 and 27 % in the hull lignin content (table 4). The variety Lutz had the highest lignin content of the hull. However, beneath the domestic conventional oat varieties two varieties (Buggy, Carron) had even a somewhat lower hull lignin content regularly (type 2). But only their rest of hydrolysis (very low detergent to acid) was different to the other domestic varieties because their acid detergent lignin content was even (3 %). No interaction between hull structure and hull colour have been detected inside all conventional oat varieties. Another type (type 3) has been represented by the breeding line 07/1422 that have been developed from a cross with AC Assiniboia. Its acid detergent hull lignin content was somewhat higher (1-1.5 %) whereas its rest of hydrolysis was remarkably lower. In the sum its hull lignin content was about 18-21 % (intermediate between conventional oats and AC Assiniboia), its hull colour was light brown. The last type (type 4) was fixed i.e. in the breeding line 07/1425. This type was the most promising in the sense of the aim of research. Its acid detergent hull lignin content was somewhat higher again whereas the more heavy acid soluble substances of the hull could be lowered further. So its hull lignin content was about 50 % of the conventional milling oat varieties only (table 4) and even somewhat lower than the hull lignin content of the donor variety. The hull colour of this type was brown (like the donor) only.

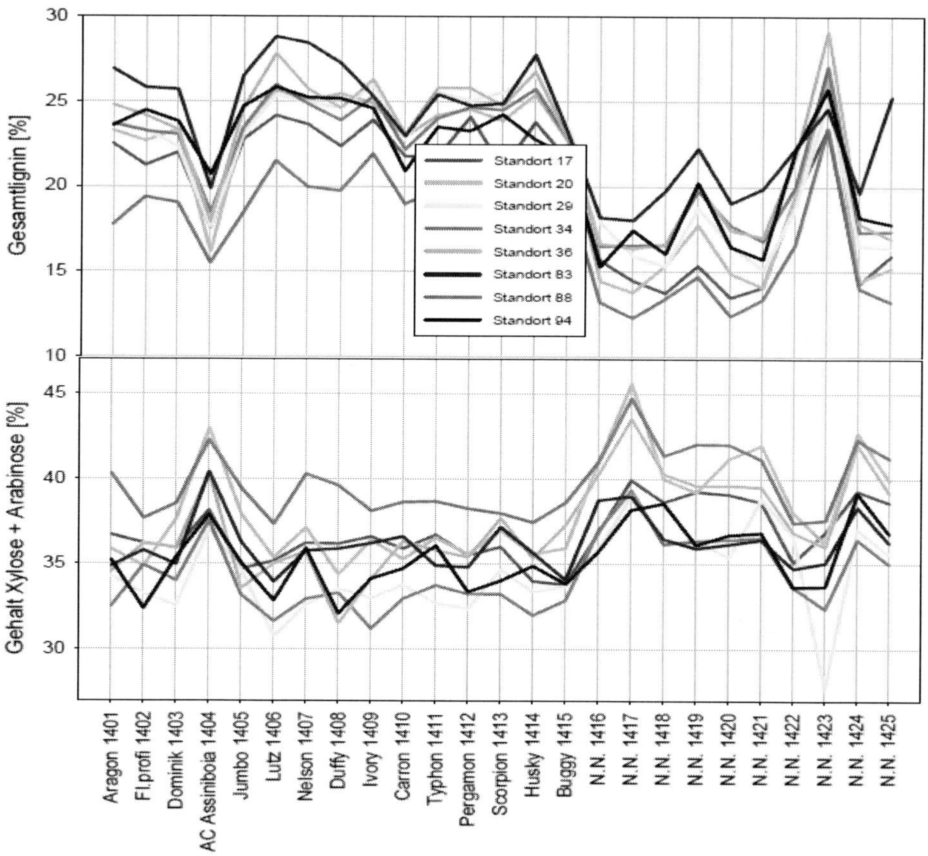

Figure 2. Hull lignin (above) and arabinoxylan (below) content 2008.

Table 4. Genetic variability of the hull lignin content 2007-2009

	Type 1 (Lutz, Ivory)	Type 2 (Buggy, Carron)	Type 3 (07/1422)	Type 4 (07/1425)	AC Assiniboia
ADL-content*	3 %	3 %	4-4,5 %	5-5,5 %	5 %
Rest of hydrolysis	20-23 %	18-19 %	13-18 %	10-13 %	12-14 %
∑ Lignin	23-27 %	21-23 %	18-21 %	16-18 %	17-19 %

*ADL = acid detergent lignin.

Up to now this kind of variability of the oat hull lignin content has not yet reported. Genotypic differences in the hull lignin content of oats already have been given before (9) but the single data can not be compared completely due to different investigation methods and a focus to the oat feeding quality.

Table 5. Genetic variability of the hull sugar content 2007-2009

	Type 1 (Lutz, Ivory)	Type 2 (Buggy, Carron)	Type 3 (07/1422)	Type 4 (07/1425)	AC Assiniboia
Xylose	28-33 %	29-32 %	29-35 %	30-38 %	31-36 %
Arabinose	2.5-3.4 %	2.7-3.4 %	3.2-3.6 %	3.5-4.1 %	3.3-3.7 %
Glucose	31-34 %	33-35 %	32-36 %	34-38 %	34-35 %
∑ Sugar	64-71 %	67-72 %	67-75 %	71-78 %	70-76 %

Not surprisingly the variability of the hull lignin content was coupled to changes in the content of further hull polysaccharides (table 5). The lower hull lignin content of the varieties Buggy and Carron (type 2) had no effect to the hull arabinoxylan content. Only their hull glucose content has been slightly increased in comparison with the other conventional milling varieties followed by a somewhat higher overall hull sugar content. The intermediate type (type 3), represented by the light brown coloured line 07/1422, combined a slightly higher amount of hull glucose with a higher hull xylan content. The fourth type (genotypes 07/1425 or AC Assiniboia) had once again even higher amounts of hull arabinoxylan and hull glucose in comparison with type 3 and consequently represented the best basis for the further development of the research.

The calibration of the NIRS method to the polysaccharides of the oat hull mostly has been a success. Especially the hull lignin content ($R^2 = 0.98$) but also the hull arabinoxylan content ($R^2 = 0.85$) could be detected on a wide range with high efficiency in future oat breeding cycles. Other authors (16,17) adapted this technology to practical oat breeding too having in mind a better digestibility of the oat hull fibre.

Conclusion

Arabinoxylan of oat hulls offers the possibility for a higher value marketing of the hulls for different technical purposes. For this reason there is a demand for adapted oat varieties with a lower lignin hull content in spite of its variability in present milling oat varieties. Special donors (like the Canadian variety AC Assiniboia) can be used.

The look for low lignin/high xylan oats with better processing ability can be successful. Due to the interaction of a bad variety performance (i.e. caused by drought) with a better processing ability a genetic improvement of the oat hulls is essential. The final assessment of the detected hull consistence for technical purposes has to be done by partners from the chemical industry and include investigations on the molecular weight of the hull arabinoxylan. Presently from a breeders point of view screening for higher yield and better agronomy is still necessary to release fully competitive xylan oat varieties in the near future. Variation in the hull consistence over years, locations and varieties exist and can be used to higher the output of the whole process. The interaction of a low lignin/high xylan oat hull with the hull colour (brown or light-brown) could be negative in terms of the future market success because there is farmer preference for non-brown oat hull varieties in the different European oat cropping regions (2). Therefore this kind of interaction should be broken in future breeding cycles supported by i.e. the successfully established NIRS-technology. The question for the utilisation of the process residue (40 %) still remains open.

Acknowledgments

This research has been funded by the German Federal Agency for Renewable Resources (FNR, FKZ 06NR066-068) under the supervision of the German Federal Ministry for Food, Agriculture and Consumer Protection (BMELV).

References

[1] Funke, C: Hafer, *Avena sativa L. - Bemühungen um das ehemals wichtigste Futtergetreide.* In: Röbbelen G. (ed.), Die Entwicklung der Pflanzenzüchtung in Deutschland (1908-2008), 100 Jahre GFP e.V. - eine Dokumentation. Vorträge für Pflanzenzüchtung 75 (2008), 312-320.

[2] Beuch, S: Züchtung auf Ertrag und Qualität bei Hafer (Avena sativa L.) – Entwicklung und Perspektiven. 61. *Tagung der Vereinigung der Pflanzenzüchter und Saatgutkaufleute Österreichs* (2010), 57-63.

[3] Brandau, J: *Supply and Demand.* Abstracts for the 2006 American Oat Workers' Conference. Fargo, North Dakota, USA (2006), p. 9.

[4] Yu, J; Beuch, S.; Herrmann, M.; Hackauf, B.: AB-QTL analysis for ß-glucan content in oats. In: Peltonen-Sainio P., Topi-Hulmi M. (Eds.), Proc. 7th Int. Oat Conf., MTT Agrifood Res. Finland, Jokioinen. *Agrifood Research Reports 51* (2004), p. 52.

[5] Cervantes-Martinez, C. T.; Frey, K. J.; White, P. J.; Wesenberg, D. M.; Holland, J. B.: Correlated Responses to Selection for Greater ß-Glucan Content in Two Oat Populations. *Crop Sci.* 42 (2002), 730-738.

[6] Andon, M. B.; Anderson, J. W.: The Oatmeal-Cholesterol Connection: 10 Years Later. *American Journal of Lifestyle Medicine* 2/1 (2008), 51-57.

[7] Puls, J.; Saake, B.: Isolierung und Derivatisierung von Hemicellulosen aus Haferspelzen und anderen Einjahrespflanzen. *FNR Final Report* 22009700 (2004).

[8] Kahlke, D.: *Xylan from oat spelts: Extraction and possible use*. In: Peltonen-Sainio P., Topi-Hulmi M. (Eds.), Proc. 7th Int. Oat Conf., MTT Agrifood Res. Finland, Jokioinen. *Agrifood Research Reports* 51 (2004), p. 105.

[9] Thompson, R. K.; Mustafa, A. F.; Mc Kinnon, J. J.; Maenz, D.; Rossnagel, B.: Genotypic differences in chemical composition and ruminal degradability of oat hulls. *Can. J. Anim. Sci.* 80 (2000), 377-379.

[10] Rossnagel, B.; Eckstein, P.; Williams, S.; Arganosa, G.; Kibite, S.; Scoles, G.: Low acid detergent lignin oat hull: molecular marker development and chromosome location. In: Peltonen-Sainio P., Topi-Hulmi M. (Eds.), Proc. 7th Int. Oat Conf., MTT Agrifood Res. Finland, Jokioinen. *Agrifood Research Reports* 51 (2004), p. 153.

[11] BSA: *Richtlinien für die Durchführung von landwirtschaftlichen Wertprüfungen und Sortenversuchen*. ed. Bundessortenamt Hannover and Biologische Bundesanstalt Braunschweig (2001), ISSN 1431-1089.

[12] Utz, H. F.: *PLABSTAT - A computer program for statistical analysis of plant breeding experiments*. University of Hohenheim, Institute of Plant Breeding, Seed Science and Population Genetics (2005), http://www.uni-hohenheim.de/~ipspwww/soft.html

[13] Sinner, M., Simatupang, M. H.; Dietrichs, H.H.: Automated quantitative-analysis of wood carbohydrates by borate complex ion-exchange chromatography. *Wood Sci. Technol.* 9 (1975), 307-322.

[14] Sinner, M.; Puls, J.: Non-corrosive dye reagent for detection of reducing sugars in borate complex ion-exchange chromatography. *J. Chrom.* 156 (1978),197-204.

[15] Willfor, S.; Pranovich, A.; Tamminen, T.; Puls, J.; Laine, C.; Suurnakki, A.; Saake, B.; Uotila, K.; Simolin, H.; Hemming, J.; Holmbom, B.: Carbohydrate analysis of plant materials with uronic acid-containing polysaccharides - A comparison between different hydrolysis and subsequent chromatographic analytical techniques. *Ind. Crop. Prod.* 29 (2009), 571-580.

[16] Lemons e Silva, C. F.; Kothe Milach, S. C.; Delmar dos Anjos e Silva, S.; Federizzi, L. C.; Montero, C. R.; Serena Fontaneli, R.: Oat fiber fractions and their application in breeding programs. *Pesq. agropec. bras., Brasília,* 41/6 (2006), 975-980.

[17] Redaelli, R.; Berardo, N.: Prediction of fibre components in oat hulls by near infrared reflectance spectroscopy. *J. Sci. Food. Agric.* 87 (2007), 580-585.

In: Organic Chemistry, Biochemistry, Biotechnology ... ISBN: 978-1-62081-155-9
Editors: G. Zaikov, F. Pudel and G. Spychalski © 2013 Nova Science Publishers, Inc.

Chapter V

Application of the Ultrasound Treatment of Flax Fibers to Produce Yarns by Open-End and Wet Flax Spinning

Zdzisław Czaplicki, Jerzy Mankowski and Waldemar Cierpucha
Institute of Natural Fibres and Medicinal Plants,
Poznan, Poland

Abstract

In this study the influence of ultrasound treatment of flax fibers (noils) to improve their quality parameters was carried-out. The assessment of the influence of fibers process was based on the spinning specimen. Two directions of the application of flax noils after ultrasonic treatment were considered. The first direction concerns the rotor spinning (OE). This direction allows the manufacturing of the cotton-blended yarns in the share of 30% flax noils (as cottoning) and 70% of cotton. In the second direction the yarns are manufactured by wet spinning flax system. The flax yarns of linear mass equal to 100 and 110 tex were obtained. The rotor-spun blended yarns produced in the process were characterized with good quality parameters. In the case of the evaluation of pure flax yarns manufactured by the wet flax system no significant differences were observed. The flax yarns made from the fibers without and after ultrasound treatment have similar parameters.

Keywords: Cotton fibres, flax noils, ultrasonic treatment, cottonin flax, blended yarns, rotor spinning, flax yarns, wet flax spinning

1. Introduction

The flax fibres are the most famous and popular among the bast fibres. They occur in the stem in the form of elementary fibres joined together by glue plant (pectin), in technical fibres. In the various phases of technological processing into the yarn the technical fibres are subject to sharing the finer bands of filaments. Dividing ability of these fibres during technological processes and the type of these processes affects the application of fibre to the proper use as the yarn or textiles. The quality of flax fibres depends on the genetic properties of plants and development of plants, depending on growing conditions and ways of extracting fibre from straw (wetting, turndown service, decortications, steaming, boiling). Research conducted for many years concern more efficient and economical use of flax fibre in both the traditional flax spinning technology as well as cotton and wool systems in blends with cotton or man-made fibres. The processing of flax fibres into yarn by other techniques requires special finishes (enriches) mainly shortening and thinning of the fibres. For the processing of flax fibre, different methods are used, for example based on chemical treatment (in the form of fibres or roving) or enzyme. The ultrasound technique has been known for a long time.

Application of ultrasounds in the technique is known from long time. Worldwide research into the application of ultrasounds in the textile industry concern among other, chemical treatment during finishing of flax fibres [1], enzymatic treatment of cotton [2,3], textile dyeing [4].

In Poland, the textile industry applies the ultrasound for many years, for example, for cutting, welding, etc. [5]. The first attempts for ultrasonic wet textile processes were conducted at the Institute for Sustainable Technologies [6] and the Institute of Natural Fibres and Medicinal Plants [7, 8]. Developed technology included the washing of the small batches of wool by using ultrasonic generators [8]. Recent studies conducted in the INF&MP relate to improvement of the spinning properties of flax fibres as a result of treatment with ultrasound [9].

2. Aim and Background

The task of the investigation is the effect of ultrasonic treatment of flax fibers (flax noils) on the improvement their quality parameters. The assessment of the influence of fibers process will be carried-out on the basis of the quality parameters of manufactured yarns. Two directions in spinning process of flax noil were assumed. The first direction assumes the production of blended yarns by open-end spinning system. As the components of a mixture it was assumed of 30% of flax noils (as cottoning) and 70% of cotton. It was assumed the manufacturing of the yarns of linear mass 40, 50, 60 and 80 tex. The second direction - the traditional spinning of flax fibers by wet spinning flax system. The flax noils before treatment and after treatment in the ultrasound will be manufactured with the yarns linear mass 100 and 110 tex.

3. Experimental and Results (Part 1) – Open-End Spinning

The studied group consisted of flax fibres in the form of biological (dew retted flax) flax noils (Ns12) and the Uzbek cotton of linear mass 0.162 tex mass, classifier length 33-34 mm, and 3.2% total impurities. The flax fibre in the form of flax noils was prepared in two variants. In the first variant the fibres are allocated to the biological treatment into the cottonin and spinning in a blend with cotton. In second variant the flax fibres before being processed as described above under identical conditions were processed in the ultrasound.

Table 1. The results of the investigations of the parameters of flax fibres before and after treatment

Parameter	Unit	Flax fibre	
		Before treatment US	After treatment US
Mean length of fibre	mm	198	193
Linear mass	tex	3,45	2,98
Fibre impurity	%	0,86	0,49

The treatment of flax fibres (biological noils) in the ultrasound was performed in ECCO Gleinftechnik GmbH (Germany) in the semi-technical scale. The results of the basic parameters of flax fibres before and after the treatment are shown in Table 1.

3.1. Cottonin Flax

The flax fibre (in the form of noils, Table 1) also was prepared in two variants to obtain the flax cottonin. The production of flax noils cottonin conducted at the plant in Steszew Division of INF&MP on the CR device. Flax noils were processed on the CR unit twice. As a result of flax noils treatment the mechanical noils cottonin: noil without ultrasound treatment - 79% of noil after treatment with ultrasound - 89.6%. The results of the investigations of basic parameters of flax cottonin are shown in Table 2.

Table 2. Results of laboratory tests of flax cottonin

Parameter	Unit	Mechanical cottonin from the flax noils	
		Before treatment US	After treatment US
Mean length of fibre	mm	34,5	29,5
Linear mass	tex	1,93	1,22
Fibre impurity	%	0,05	0,03

The results indicate that better quantitative and qualitative parameters of the treatment of noils on the device CR were achieved for fibres with the use of ultrasound treatment of noils.

3.2. The Blend of Flax Cottonin and Cotton

The mixing of flax cottonin with the cotton as well as the preparation of card was performed on the machine park in the plants 'Bonitex' in Boleslawiec. The blend flax - cotton was made in the proportion of 30% flax cottonin and 70% cotton. The flax cottonin was moistened with liquid K dissolved in water at the proportion 1:10, before making a blend. The share of liquid K to the mass of flax cottonin was 0.3%. Flax cottonin and cotton was spread in layers by hand in a climate chamber. The time of seasoning was 4 hours. After mixing the fibres the blend was transported by a pneumatic way to the carding machine OCTIR made in Italy. This double drum-carding machine is equipped with 9 solid caps, and is fitted with a device, which uses electrical adjustment and delivers the band into two cans. The carding for both variants (with the use in noils US and without) proceeded in a proper manner. As result, the card sliver, which contains the parameters given in Table 3, was obtained.

Table 3. Quality parameters of the card sliver

Parameter	Unit	Cottonin card sliver	
		Before treatment US	After treatment US
Linear mass	tex	4045	4021
Variation coefficient of linear mass	CV %	2,75	2,45

3.3. Open-End Spinning

The card sliver from a blend of flax - cotton made in the plant "Bonitex in Boleslawiec (Poland) has been subjected to the spinning on the rotor spinning machine (BT905 BASEtex) at the rotor speed 40000 rpm. In the spinning mill were applied the following environmental conditions: temperature19.6 0C and relative humidity 58%.

The yarns were produced from the fibres, both US-treated and untreated with linear mass: 40, 50, 60 and 80 tex and twists respectively: 1100, 1000, 900 and 750 twists per meter (T.P.M.).

3.4. Results and Discussion

The laboratory tests included the measurements of manufactured yarns: linear density, tenacity, linear density irregularity and errors (Uster).

The results of the investigations of yarn parameters are shown in Table 4 and Figures 1-6. The results obtained in treatment of flax noils with ultrasonic treatment as compared to the untreated noils allow formulating the following statements and remarks:

- flax noils treated with ultrasound (US) in the phase before being processed, have underwent minor changes in terms of fibre length, linear density and the level of impurities,

- flax noils after treatment in the ultrasound and used in the production of cottonin provide a greater yield (yield) of cottonin (89.6% vs. 79.0%) and were significantly thinner to 1.22 vs. 1.93 tex.
- rotor (OE) cotton yarns that were obtained with 30% flax cottonin and after the treatment with ultrasound have favourable parameters such as coefficient of variation of linear density, the number of thinning and swells.
- tenacity of yarns appropriate for comparable variants of spinning is generally comparable with a slight majority in favour of the flax noils yarns without ultrasonic treatment, while the coefficient of tensile strength variation is usually lower than what may be a slight weakening of the fibres after treatment US.

Table 4. Quality parameters of blended yarns (70% cotton, 30% cottonin flax)

Linear mass of yarn tex	nom.	40		50		60		80	
	real	40,1	41,4	45,5	50,9	61,6	64,2	74,68	84,3
Type of specimen without treatment, after treatment US		Before treat-ment	After US	Before treat-ment	After US	Before treat-ment	After US	Before treat-ment	After US
Tenacity cN/tex		8,92	9,12	9,32	9,21	9,53	9,2	10,1	9,82
Coefficient of tencile strength variation %		11,55	8,62	7,96	8,71	7,03	10,0	11,4	6,95
Breaking elongation %		9,31	9,46	9,51	9,59	10,0	9,9	9,89	10,2
Coefficient of the variation of linear mass %		17,5	16,7	17,4	16,2	16,2	15,4	16,0	14,5
Number of thinning (-50%) on 1 km		24	9	17	6	5	3	4	0
Number of swells (+50%) in 1 km		149	92	150	79	112	55	77	36
Number of swells (+100%) in 1 km		1	0	1	0	1	0	0	0

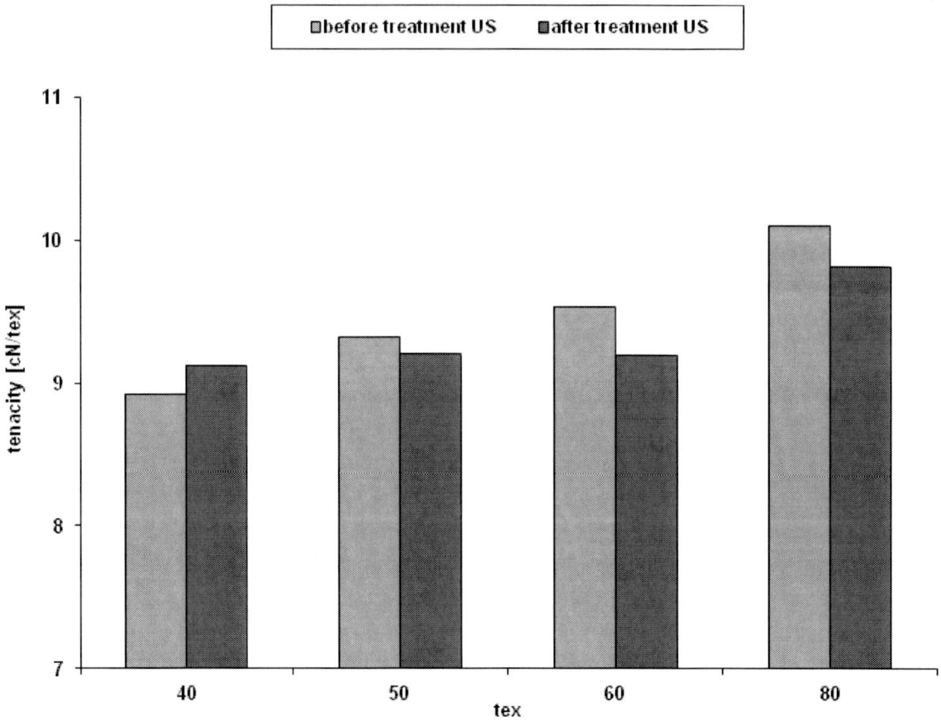

Figure 1. Tenacity of yarn for different linear mass (tex) made from blended cotton fibres and flax.

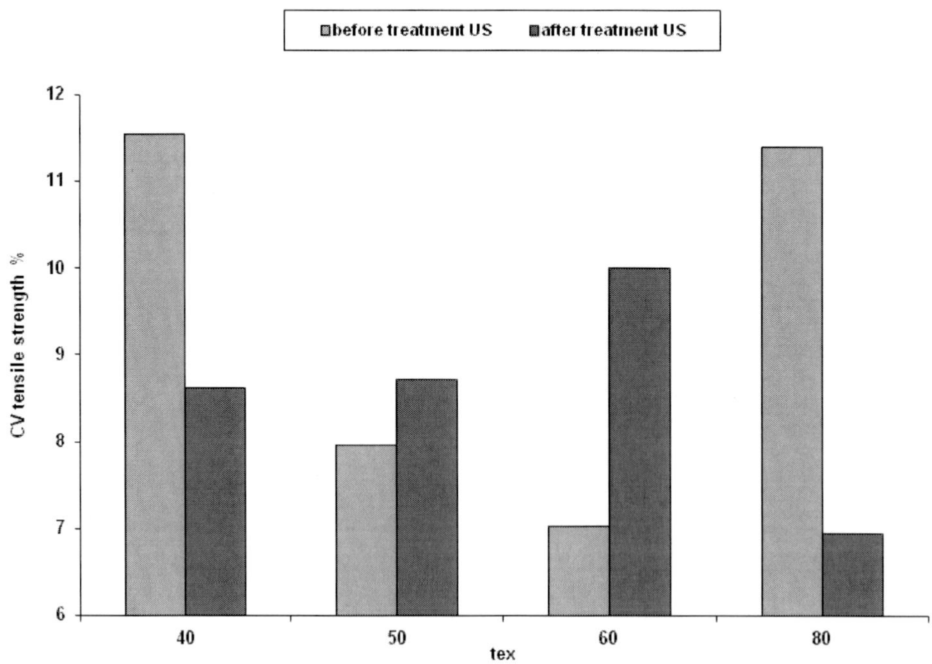

Figure 2. Coefficient of tensile strength variation for different linear mass yarns.

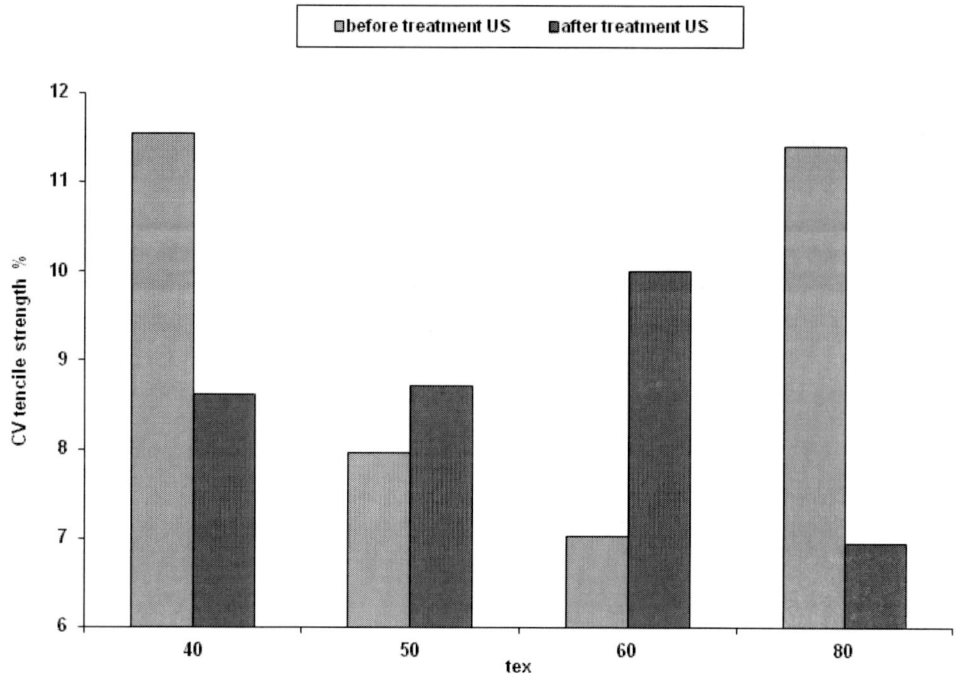

Figure 3. Breaking elongation for different linear mass yarns.

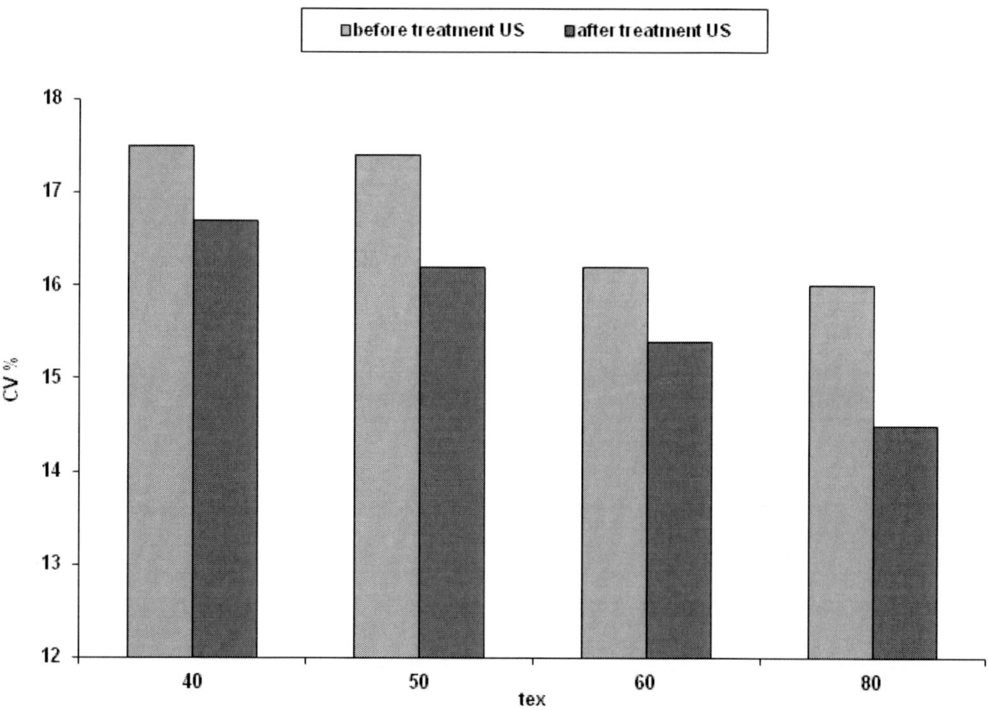

Figure 4. Coefficient of the variation of linear mass for different yarns.

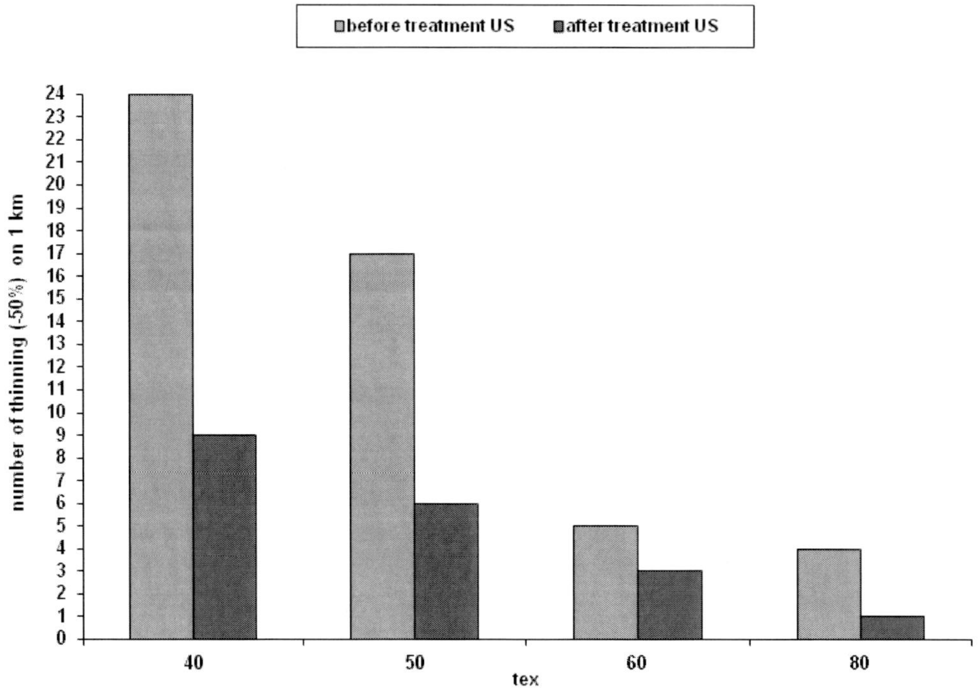

Figure 5. Number of thinning (-50%) on 1 km for different yarns.

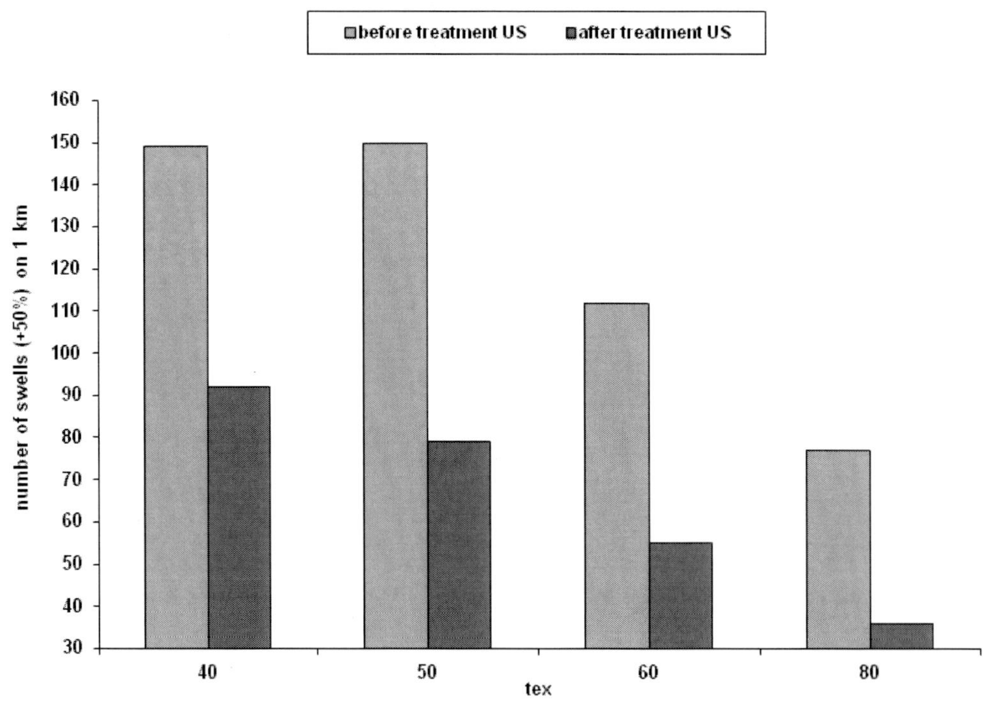

Figure 6. Number of swells (+50%) on 1 km for different yarns.

4. Experimental and Results (Part 2) - A Wet Flax Spinning System

4.1. Method of Preparation of Flax Noils for the Spinning by the Wet Spinning System

Two similar samples of biological flax noils Ns14 (dew retted flax) of similar weight were prepared for the testing. One sample was prepared for direct flax fiber wet spinning system and the second sample before the spinning under identical conditions was processed in the ultrasound. Table 5 shows the fiber quality parameters before and after treatment in the U.S..

Table 5. Quality parameters of flax noils before and after treatment

Parameter	Unit	Flax fibre (noils)	
		Before treatment US	After treatment US
Mean length of fibre	mm	205.1	197.0
Linear mass	tex	2.85	2.45
Fibre impurity	%	0.55	0.25

4.2. Spinning of Flax Noils by Flax Wet System

In the spinning tests were used (dew retted flax) the flax noils (Ns14) two samples - before and after ultrasound treatment - Table 5.

The spinning was carried out in Flax Plant "Madex" in Malbork (Poland) in the production conditions, using the following machinery and equipment:

1) Carding machine Gz600L
2) GN-type drawing frame - supplied by six carding slivers
3) Drawing frame 1 (made of Metalen) - powered by 4 slivers (for one outlet)
4) GN-type drawing frame – supplied by 6 slivers on the first drawing frame
5) Draw frames 2, 3 and 4 (made of Metalen) - supplied by 4 slivers (for one outlet)
6) Roving frame J. Mackie (Belfast)
7) PM-88Ł3 spinning machine (for 100 tex) and PM-114 (for 110 tex)
8) Dryer
9) Winder Autoconer 338

In the carding process the output of carding sliver made of fibers without the use of ultrasound was equal to 85.1% however after the application of ultrasound the output was 88.1%. Table 6 gives the fiber quality parameters in the carding sliver and after the drawing frame GN.

Table 6. Quality parameters fibre before and after treatment US

Parameter	Unit	Fibre before treatment US		Fibre after treatment US	
		Carding sliver	Drawing sliver (after GN)	Carding sliver	Drawing sliver (after GN)
Mean length of fibre	mm	190.2	178.0	157.3	152.0
Linear mass	tex	2.46	2.35	2.2	2.05
Fibre impurity	%	0	0	0	0

Quality parameters of fiber flax yarns before and after treatment in ultrasound arc shown in Table 7.

Table 7. Quality parameters flax yarns (wet spinning)

Linear mass of yarn tex	nom.	100		110	
	real	100	95.2	112.5	108.5
Type of specimen before treatment and after treatment US		Before treatment	After US	Before treatment	After US
Coefficient of the variation of linear mass %		6.25	6.62	5.95	6.23
Tenacity cN/tex		14.2	14.3	14.1	13.9
Coefficient of tensile strength variation %		22.8	18.6	21.8	23.2

4.3. The Evaluation of the Spinning and Research Results

The spinning process was normal with no disruption at all stages, i.e. carding, drawing, roving and spinning. In the organoleptic assessment of fibers after treatment with ultrasound as well as the half products and yarn were characterized by a significantly lighter shade and higher degree of delicacy. At the same time, a higher susceptibility to the shortening of fibers and a greater ability to share into the groups of thinner fibers was observed.

In the process of carding the fibers after ultrasonic a greater efficiency of carding sliver by about 3%. In the process of spinning the ends down of yarn was within the limits adopted in the standards.

The comparison of the quality parameters of yarns made from flax fibers before treatment and after treatment in the ultrasound revealed no significant differences.

Conclusion

In all phases of investigations the correctness of process during both the preparation of half products and spinning was confirmed. This applies both to the production of flax cottonin, blending with cotton and Open-End Spinning.

1) The results have shown that the fibre in flax noils after the ultrasound treatment and according to the organoleptic evaluation are clearly of lighter shade, and are softer, more delicate but slightly attenuated. These fibres have a greater ability to share technical fibres in the processes of mechanical treatment, especially in terms of cottoning (intense shortening and thinning of fibre).
2) The use of flax noil cottonin after treatment US in cotton Open-End Spinning system in a blend with cotton (70%) had a positive impact on the course and results of spinning. Yarns were obtained with higher quality parameters. This is due to significant thinning of cottonin (to 1.22 tex) in the re-treatment of noils and after application of ultrasound, which obviously increased the number of fibres in the yarn cross-section.
3) Application flax noils after ultrasound treatment in the wet flax spinning system have showed no influence on the improvement of the quality parameters of yarns.

References

[1] Drozdow W.G., Komow A.A., Lisku W.G.: *Wlianie ultrazwukowych kolebanij na process chemiceskowo oblagorarziwania lnianego łuba.* Technologija Tekstylnoj Promyszlenosci No.5, 1976.

[2] Yachmener G. at al.: *Use of Ultrasonic Energy in the Enzymatic Treatment of Cotton Fabric.* Southern Regional Research Centre, New Orleans, Louisiana 1998.

[3] Yachmener G. at al.: *Study of the Influence of Ultrasound on Enzymatic Treatment of Cotton Fabric.* US Department of Agriculture, Southern Regional Research Centre, New Orleans, Louisiana, September 1999.

[4] Vajnhandl S., Majcen Le Marechal A.: *Ultrasound in textile dyeing and the decolouration / mineralization of textile dyes.* Dyes and Pigments, vol. 65. Issue 2, May 2005.

[5] Kozłowski R., Ruszkowski K., Czaplicki Z.: *Application of ultrasounds in the textile industry.* Textile Overview WOS, nr. 8, 2006. (in Polish)

[6] Ruszkowski K.: *Dyeing of textiles with the use of ultrasounds.* Textile Overview WOS, nr. 7, 2007. (in Polish)

[7] Patent P-381 860. *Method and device for the intensification of wet textile processes.* Patent Office of RP, 28 February 2007.

[8] Czaplicki Z., Ruszkowski K.: *Technology of the washing of wool small parties with the use of ultrasounds generators* Textile Overview WOS, nr. 1, 2010. (in Polish)

[9] Mańkowski J., Cierpucha W., Czaplicki Z., Kołodziej J. i Mańkowski T.: *The rotor blended yarns made of the cotton and flax cottonin after the ultrasound treatment.* Textile Overview WOS, nr. 4, 2011. (in Polish)

In: Organic Chemistry, Biochemistry, Biotechnology ... ISBN: 978-1-62081-155-9
Editors: G. Zaikov, F. Pudel and G. Spychalski © 2013 Nova Science Publishers, Inc.

Chapter VI

Sustainable Biofuels: An International Overview of Existing Approaches

Joerg Kretzschmar
Deutsches Biomasse Forschungs Zentrum gGmbH (DBFZ), Leipzig, Germany

Abstract

The directive 2009/28/EC among others defines different criteria for sustainable biofuel production. The main objective in this context is the improvement of the GHG balance of transport fuels. Considering the three-pillar-system of sustainability that aims to reconcile environmental, social and economic demands it is clear that a sustainable production of biofuels has to include much more criteria than given in the directive. Within the scope of this study, an indicator based research on sustainable biofuel options was done on international level.

Keywords: Biofuels, sustainability, wastes, residues, national renewable energy action plan

Introduction

Biofuels have to be produced in a sustainable way. Therefore, article 17 of the directive 2009/28/EC defines the following criteria. Biofuels have to save at least 35% greenhouse gases (50 % in 2017 and 60 % if production started at or after the 1st January 2018) compared to the fossil reference (83.8 gCO_{2eq}/MJ according to the directive).

To avoid negative land use changes, the raw materials for biofuel production shall not be produced on land with high carbon stock such as wetlands, continuously forested areas and peatland. All of the above mentioned criteria consider only direct land use change, while indirect land use change, where areas with high carbon stock serve as new arable land for those crops that had to give way for energy crops on the initially cultivated land, is not regarded. To secure biodiversity, production shall not take place on land such as primary forest or other wooded land, areas designated for nature protection purposes and on highly biodiverse grassland. Agricultural raw materials cultivated in the Community and used for the production of biofuels shall be produced in accordance with the minimum requirements for good agricultural and environmental condition according to the council regulation (EC) No. 73/2009. The directive 2009/28/EC does not define social criteria which are directly connected to the production of raw materials.

Social sustainability is just regarded via a report duty of the Commission to the European Parliament and the Council including the availability of food, the respect of land-use rights and the ratification of different core conventions of the International Labour Organisation (ILO). Therefore, the main gaps are to be found in missing criteria to avoid indirect land use change and specific social criteria that have to be included in biofuel certification schemes. If the future biofuel production really wants to contribute towards the climate change, it has to consider all relevant sustainability criteria, independent of existing directives and regulations.

The sustainability problems of biofuel production are mostly connected to land intensive 1st generation biofuels. Biofuels from waste and residues do not directly compete with agricultural production and enable the cascade use of biomass (material use before energetic utilisation) so that they are more likely to be designated as sustainable, especially if international human rights and labour standards are considered during the production process.

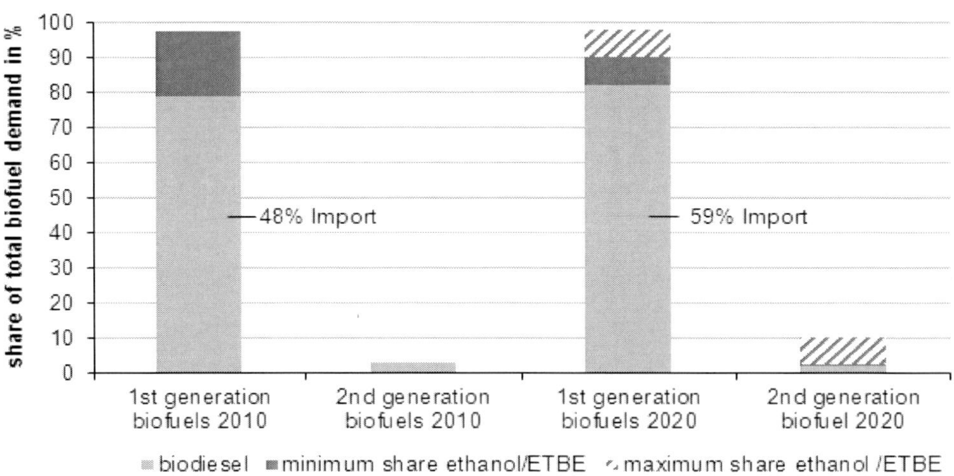

Figure 1. composition of total biofuel demand in 2010 (144 PJ) and 2020 (222 PJ) according to the German National Renewable Energy Action Plan 2010 [1].

Background

According to Article 4 2009/28/EC, every Member State had to submit general national renewable energy action plans (NREAP) by June 2010. These plans include, among others, goals concerning biofuel ratios to reach the required GHG emission reduction within the transport sector in 2020. The German NREAP 2010 for example shows that the majority of biofuels used in 2020 will be 1st generation biofuels (see Figure 1). In 2010 about 3 % and in 2020 only 2.49 – 10.19 % of the total biofuel demand will be 2nd generation biofuels. 59 % of the required 1st generation biofuels will be imported in 2020. With view to a sustainable biofuel provision, this initial situation shows the need to develop the biofuel certification with regard to indirect land use change and social sustainability (e.g. living and working conditions, salaries, market access of smallholders) on the one hand and to increase the share of 2nd generation biofuels on the other hand.

It was the aim of the international project review to identify biofuel production concepts and techniques as well as current R&D projects which could be identified as sustainable, based on a positive evaluation of the indicators raw material, land use change, agriculture, competitive uses, GHG saving potential, stage of development, form of application and factors of social sustainability, insofar evaluable.

Table 1. Summary of biofuel projects which contain elements of a sustainable production, GHG emission reduction refers to the fossil reference according to 2009/28/EC

Company/ Project (Land)	Feedstock	Biofuel	GHG emission reduction
Enerkem (CA) [2]	sorted domestic waste, waste wood	ethanol from syngas (BtL)	87 %**
Verbio (DE) [3]	grain, rapeseed, vinasse from ethanol production	biodiesel/ ethanol biomethane	38 %**/ 81 %*/ ---
Sekab (SE) [4,5]	certified sugarcane (own certification system)	ethanol	71 %**
Neste Oil (FI) [6]	palm oil, rape oil, waste oil	NexBtL (HVO)	~60 %***
St1 Biofuels Oy (FI) [7]	food processing residues	ethanol	80 %*
Biogaspartner/ BRV GmbH (DE) [8,9]	catering waste and decayed food	biomethane	73 %**
Cleantech Biofuels (US) [10]	domestic waste	only feedstock production	---
Magne Gas (US) [11]	Sludge, waste water, in general fluid waste with up to 10 % dry substance	MagneGas™ (CMG)	---
Aquafinca Saint Peter Fish (HN) [12]	fish oil	biodiesel	83 %
Seambiotic (IL/US) [13]	alga and flue gas	ethanol/ biodiesel	---
Niwa (NZ) [14]	alga and flue gas	bio-crude	---
Ineos Bio (US/GB) [15]	organic waste and residues	ethanol	90 %*
Joule (US) [16]	CO_2, water, nutrients (DirectFuel technology)	ethanol, diesel	90 %*
Etamax (DE) [17]	lignin low organic waste and alga	biomethane	73 %**
Bioliq (DE) [18]	straw, hay, waste wood	BtL fuel	95 %**

--- no data available, *manufacturers data, **BLE Default Values [19] , ***studies.

Experimental Part / Results

After a first screening, projects which possess a high potential to provide future sustainable biofuels were identified via a qualitative analysis of the chosen indicators as well as the conversion technologies.

By this means, an insight into different international approaches towards the production of sustainable biofuels was obtained. The indicator based research resulted in 15 projects which include elements of a sustainable production. In Table 1 all 15 projects are summarised. They have to be understood as a first overview. Due to the consistent consideration of all indicators, only six biofuel options remained. Decision support was given among others by the matrix in Table 2.

Table 2. Decision matrix

Project	Food vs fuel	LUC	GHG saving potential	Stage of development	Niche application	Social sustainability
Enerkem	No	No	High	High	No	n.e.
Verbio	Yes	No	High	High	No	n.e.
Sekab	Yes	n.e.	High	High	No	Yes
Neste Oil	Yes	n.e.	High	High	No	n.e.
St1 Biofuels Oy	No	No	High	High	No	n.e.
Biogaspartner	No	No	High	High	No	n.e.
Cleantech Biofuels	No	No	n.e.	High	No	n.e.
Magne Gas	No	No	n.e.	High	Yes	n.e.
Aquafinca	No	No	High	High	No	n.e.
Seambiotic	No	No	n.e.	Low	n.e.	n.e.
Niwa	No	No	n.e.	Low	Yes	n.e.
Ineos Bio	No	No	High	Low	n.e.	n.e.
Joule	No	No	High	Low	n.e.	n.e.
Etamax	No	No	High	Medium	n.e.	n.e.
Bioliq	No	No	High	Medium	n.e.	n.e.

n.e.: not (yet) evaluable.

All projects of interest exclusively use waste and byproducts as raw material and therefore are not directly connected to any sustainability criteria in the field of agriculture like direct or indirect land use change (LUC/ILUC) or competition with the food production. The applied conversion technologies range from gasification and fuel synthesis over anaerobic fermentation followed by distillation and/or rectification to biomethane production. The results are summarised in Table 3.

Discussion

The negative sustainability of biofuels is first of all connected to land intensive first generation biofuels. However, actual climate protection goals lead to a high demand of biofuels which can only be covered by using first generation biofuels. Because of that high demand, even food based biofuels can be found in Table 1. Strictly speaking, they are only sustainable until the competition with food production leads to higher prices or food scarcity. The 1st generation biofuels in Table 1 are characterised by effective biomass utilisation (e.g. biomethane production from ethanol production vinasse, Verbio) or sustainable provision of feedstock. SEKAB ensures the social sustainable production of sugar cane by the use of their own certification system. The Finnish company Neste Oil only uses palm oil from land that was not rainforest or other wooded land before November 2005. The membership of the Roundtable on Sustainable Palm Oil furthermore shows active efforts to enhance the sustainable production of palm oil. Even though these certification systems do not yet account for indirect land use change (ILUC) and have only little reference to indicators of social sustainability, they mark first important steps towards sustainable biofuel production. They have to be continuously developed and improved in the future.

Due to the consistent consideration of all chosen indicators, only six sustainable projects remain in Table 3. All other projects use food as feedstock, describe niche applications, are in early developmental stage or simply produce just feedstock for biofuel production, as described in the case of Cleantech Biofuels. As already mentioned, all projects in Table 3 use wastes or residues as feedstock. The potential of this feedstock is of course not sufficient to cover the total biofuel demand, but it is the only sustainable feedstock that is available immediately. Its potential could be raised by intensification and expansion of gathering and adaption of existing biofuel conversion technologies, as already done in the mentioned projects.

Table 3. Sustainable biofuel projects identified within the study

Project	conversion technology	stage of development	capacity
Enerkem (CA) [2]	thermochemical gasification (syngas) of organic matter from domestic waste and residual wood followed by catalytic conversion to methanol, ethanol and other products	industrial	5 Mio. l/a, expansion planned
St1 Biofuels Oy (FI) [7]	decentral fermentation and distillation of ethanol, central dehydration and purification	industrial	>4,5 Mio. l/a
Biogaspartner/ BRV GmbH (DE) [8,9]	conventional production of biogas, biomethane upgrading with membranes	industrial (medium scale)	25.000 MWh/a (300 Nm^3/h)
Aquafinca Saint Peter Fish (HN) [12]	biodiesel production from fish oil	industrial (small scale)	6000 l/d
Etamax (DE) [17]	multistage biogas production	R&D	no production
Bioliq (DE) [18]	decentral flash pyrolysis followed by entrained flow gasification and Fischer-Tropsch synthesis	R&D	no production

Conclusion

1st generation biofuels could be produced more or less sustainably. For future application, existing directives have to be further developed, especially in the case of ILUC and social sustainability. To take the pressure away from land intensive first generation biofuels, other sustainable resources have to be developed or used more intensely. Suitable resources, because they are already available, are wastes and residues particularly as they enable the cascade use of biomass. In the case of waste based biofuels the problem is not necessarily the conversion technique since the results of the international review show different conversion techniques for different substrates and biofuels. Nevertheless, a lot of adaption and especially technical up-scaling has to be done. The problem is rather the low potential of and the sufficient supply with biomass for biofuel production. Therefore, 1st generation biofuels, as a first step towards sustainable transport fuels, have to be used as sustainably as possible and as little as necessary. As a logical consequence, the share of more sustainable biofuels, like biofuels from wastes and residues, has to be maximised. The potential could be raised on the one hand by a strict separation of organic fractions, as already defined within the directive 2008/98/EC and on the other hand by an enhanced acquisition of for example agricultural and silvicultural residues. A higher share of biofuels from waste in the total biofuel supply in combination with increased efficiency of combustion engines and lower fuel consumption in general, could lead in sum to sustainable goods and passenger traffic and a higher acceptance of biofuels.

Acknowledgments

The research project “Social and behavioural aspects of biofuel production and use”, as the basis for this study, was generously supported by the Federal Ministry of the Environment, Nature Conservation and Nuclear Safety (BMU).

References

[1] Bundesrepublik Deutschland: *Nationaler Aktionsplan für erneuerbare Energie gemäß der Richtlinie* 2009/28/EG zur Förderung der Nutzung von Energie aus erneuerbaren Quellen; 2010.
[2] ENERKEM - *Home, URL: http://enerkem.com/en/home.html* (accessed: 10.02.2011)
[3] VERBIO Vereinigte BioEnergie AG - *Home, URL: http://www.verbio.de/desktopdefault.aspx* (accessed: 12.03.2010).
[4] Anders Fredriksson, SEKAB: *About sustainable ethanol, URL: http://www.sustainableethanolinitiative.com/default.asp?id=1062* (accessed: 5.03.2010)
[5] SEKAB Biofuels & Chemicals AB: *Bioethanol as fuel, URL: http://www.sekab.com/default.asp?id=1887&refid=1977* (accessed: 15.03.2010)
[6] Neste Oil: *Change is a state of mind* - Annual Report 2009; Neste Oil Corporation, 2010.

[7] St1 - Biofuels Waste to ethanol, *URL: http://www.st1. eu/index.php?id=2874* (accessed: 19.04.2010).
[8] Deutsche Energie-Agentur GmbH (dena): *Biogaspartner: Kißlegg-Rahmhaus, URL: http://www.biogaspartner.de/index.php?id=12249* (accessed: 01.12.2010).
[9] Deutsche Energieagentur (dena): *Biogaspartner: Einspeiseatlas, URL: http://www. biogaspartner.de/index.php?id=10063&L=gsjykpsmyqme* (accessed: 04.10. 2010).
[10] CleanTech Biofuels, *URL: http://www.cleantechbiofuels.net* (accessed: 02.03.2010).
[11] MagneGas Corporation, *URL: http://www.magnegas.com* (accessed: 03.03.2010).
[12] Piccolo, T.: *Framework Analysis of fishwaste to biodiesel production – Aquafinca –* Case Study; Aquatic Biofuels, 2008.
[13] Seambiotic – *Home, URL: http://www.seambiotic.com/* (accessed: 12.05.2010).
[14] National Institute for Water and Atmospheric Research: *NIWA - Bio-oil from wastewater algae, URL: http://www.niwa.co.nz/our-science/freshwater/research-projects/all/biofuel-from-wastewater-algae#null* (accessed: 10.05.2010).
[15] Welcome to INEOS *Bio - Ineos Bio - clean biofuel, URL: http://www.ineosbio.com/57-Welcome_to_INEOS_Bio.htm* (accessed 04.03.2010).
[16] Spagnoli, F.: *Joule Biotechnologies: Solving the energy crisis with affordable, renewable clean fuel, URL: http://joulebio.com/* (accessed 04.03.2010).
[17] Fraunhofer Institut für Grenzflächen und Bioverfahrenstechnik: *Jahresbericht 09/10; Fraunhofer IGB*, Stuttgart, 2010.
[18] Hippler, H.; Umbach, E.: *Bioliq - Forschungszentrum Karlsruhe; URL: http://iwrwww1.fzk.de/bioliq/index.html* (accessed: 10.05.2010).
[19] Bundesanstalt für Landwirtschaft und Ernährung (BLE): *Leitfaden nachhaltige Biomasseherstellung; BLE,* Bonn, 2010.

In: Organic Chemistry, Biochemistry, Biotechnology ... ISBN: 978-1-62081-155-9
Editors: G. Zaikov, F. Pudel and G. Spychalski © 2013 Nova Science Publishers, Inc.

Chapter VII

Renewable Resources and Plant Biotechnology: Hemp Essential Oil – Production, Characterization, Application

Ryszard Kaniewski, Joanna Banach and Józef Jakimcio-Turowski
Institute of Natural Fibres and Medicinal Plants, Poznan, Poland

Abstract

Institute of Natural Fibres and Medicinal Plants in Poznan is an interdisciplinary research organization that deals with complex studies on obtaining and processing of natural textile raw materials. For many years INF&MP works on obtaining and studying essential oils from Polish hemp cultivars. The studies include breeding and agronomy of hemp and the use of by-products of its processing and production [1,2]. One of the directions of hemp use is application of essential oil in market products. The hemp essential oil/ (HEO) has important properties in terms of fragrance and biological and microbiological activity.

The predicted growth of hemp cultivation for energy production and pulp and paper will create the need to utilize hemp panicles, so far treated as useless waste product. The new technology of mechanical harvest, developed at INF&MP, makes it possible to harvest the panicles separately and then to obtain hemp essential oils by water vapour distillation.

Keywords: Hemp, hemp harvesting, essential oil, aphids, pathogens, Cannabis sativa L., Hemiptera, Alternaria

Introduction

Hemp are characterized with quite specific smell which comes from most volatile monoterpenes and sesquiterpenes. The literature data [26] reports that HEO is composed of 63,24 % monoterpenes and 31,51 % sesquiterpenes, what has been confirmed by our own studies. The main components of HEO are: α-Pinene, β-Pinene, Δ3-Carene, Myrcene, Limonene, β-Phellandrene, cis-Ocimene, trans-Ocimene, α-Terpinene, trans-α-Bergamotene, β-Caryophyllene, β-Humulene, β-Farnesene, β-Selinene, Selina-3,7(11)-diene and others [22,23,25].

Due to its characteristic flavour HEO is applied for the production of: Cosmetics, soap, creams, cosmetic oils, perfume, food products e.g. alcoholic beverages such as beer and champagne. Additionally, HEO is used in aromatherapy, also in plant protection [8,24] and in other applications [21].

Another important property of HEO is its bacteriostatic activity, mainly towards Gram+ bacteria (*Staphylococcus i Streptococcus*).

HEO is obtained from hemp panicles harvested mechanically with the use of special harvester [15].

Production of HEO

Currently the production of HEO is based on fibrous hemp varieties (Photo 1) and is run at technical scale at the Experimental Farm of INF&MP in Petkowo, Poland [6,7,9].

Preliminary trials on producing HEO were started as early as in 2002 at the Institute. The trials were made with Polish varieties of hemp (Beniko and Białobrzeskie) characterized with low content of cannabinoids, including THC (below 0,2% in dry mass). The se varieties comply with the legal requirements described in the Bill of Law of 29th July 2005 (Journal of Law no. 179, par. 1485) and are admitted for cultivation in Poland.

Photo 1. General view of the industrial hemp plantation of Białobrzeskie variety.

Harvest of Panicles

The new technology of mechanical harvesting developed at INF&MP allows for the separate harvest of panicles. The panicles practically do not contain any fibre thus transporting them together with the stems to the processing plants for fibre extraction is not economically viable. The panicles constitute significant biomass i.e. at the sowing density of 60 kg/ha 1ha yields about 7000 kg of panicles. After drying to 12% moisture content, this gives 1750 kg (according to the studies in 2002) [3,4,5,17].

Photo 2. A combine harvester for hemp.

The new technology [16] involves cutting off the panicles during harvesting the industrial hemp and transporting them on a trailer. The harvest is done during full flowering phase of hemp growth. The fresh panicles are transported directly to a distillery. When the harvest is done from the hemp cultivated for seeds i.e. at the full maturity of seed, the panicles are dried at the temperature below 40 ^{0}C and further threshed to obtain the seeds. These panicles after threshing can also be used for extracting HEO.

Regardless of the purpose of hemp cultivation (for seed or fibre) the technology enables to receive panicles. Such separate harvest of panicles will undoubtedly lead to some loss in fibre yield. This results from cutting the stems at certain height, adjusted along the work of the combine harvester (Photo 2). However the wastes produced in this system can be utilized.

Distillation of HEO

Hemp essential oils obtained from hemp panicles are extracted by water vapour distillation method. The process involves separating volatile components that are present inside the cells located near the plant surface [20]. Subsequently, these volatile substances are condensate to oil and separate aromatic water (hydrolate).

Photo 3. The distillery for essential oils.

The production is carried out with a semi-technical device (Photo 3) made of steel certified as suitable for contact with food components [6]. A single batch of fresh panicles is 2,3 m^3 , what is equivalent of about 430 kg. The process takes about 3 h and finishes with observable decrease of efficiency. Under these process conditions, in case of fibrous hemp (mechanical harvest at full flowering) the amount of 3-5 $dm^3 \cdot ha^{-1}$ of HEO can be reached.

The fresh product is subjected to further processing i.e. purification on filtration membranes and drying with sodium sulfate i.e. refining to create a commercial product.

The Analysis of HEO

The tests carried out on HEO produced from Białobrzeskie variety revealed the following characteristics of the substance.

Characteristics of HEO:

- clear liquid,
- colour from light yellow to dark green,
- smell- characteristic for hemp, forest-meadow fragrance note, pleasant,
- relative density at 20 ^{0}C – on average 0,840 g/cm^3
- light refraction index – n_D^{20} = 1.4897
- solubility in ethanol –1 ml of HEO after adding 20 ml of ethanol 95% (V/V) remains in the opacity point,

Content of essential oil in different parts of the plant such as panicles, stem and leaves varies with the highest amount present in panicles (Table 1) [14].

Table 1. Content of hemp essential oils in different parts of plant Białobrzeskie

Content of hemp essential oil [%] in:		
Panicles	Stems	Leaves
0.17	0.02	0.1

The chemical composition of the hemp essential oil extracted from Bialobrzeskie variety in 2009 presented as follows:

- 32,3 % Myrcene,
- 14,8 % β-Caryophyllene,
- 11,9 % Terpinolene,
- 10,5 % Limonene,
- 10,3 % trans-β-Ocymene,
- 7,5 % α-Pinene,
- 5,1 % α-Humulene,
- 3,2 % β-Pinene

In trace amounts (below 1 %) another 50 components.

THC content was also tested[14] in hemp essential oils from Białobrzeskie variety (yields 2008-2010). The results of these tests are presented in Table 2, where THC content in panicles is compared in different yields.

Table 2. THC content in hemp

THC content	In essential oils			
	2008	2009	2010	Średnio
mg/ml	0,401	0,116	0,086	0,020
%	0,0468	0,0140	0,0103	0,0237
Oil density at 20^0C [g/cm^3]	0,857	0,828	0,835	0,840
THC content	In panicles			
mg/ml	-	-	-	-
%	0,0601	0,0336	0,0198**	0,0378

The Application of HEO in Plant Protection

1) Aphids

The proven efficiency of HEO in fighting aphids (Dysaphis plantaginea pass.) present at apple trees led to patenting this substance for the application [27].

The aim of the study was to determine the efficiency of the HEO obtained from hemp panicles in fighting aphids (Dysaphis plantaginea Pass.) feeding on apple trees [10,13,19]. The HEO was applied at concentrations of 0,025; 0,05 and 0,1%. As the reference substance served the insecticide Mospilan 20 SP (acetamiprid) at the dose 0,125kg/ha. The treatments were carried out in July, at the developmental stage of the apple tree at BBCH, by fine drop spraying with STIHL SR 420 sprayer, using 750 l of functional liquid/ha. Each combination was repeated four times. In a single run of tests 10 long shoots infested by aphid colonies were analyzed on 4 randomly selected trees (acc. to EPPO PP 1/21 Standard). The mortality rate of the pest was determined after 24 hours and again after 5 days following the treatments. The studies allowed for observing high efficiency of HEO in fighting aphids on apple trees.

2) Spider mites

Equally high efficacy of HEO was observed in case of infestation with other plant pests e.g. spider mites. They are one of the most stubborn and difficult to combat pests of both greenhouse and field plantations. These mites attack numerous vegetable plants such as tomato, cucumber, red pepper and carrots and virtually all decorative plants, especially roses. In field plantations spider mites attack most frequently strawberries.

As the HEO proved to be effective for this application, it has been for use in plant protection against mites [28].

The patent refers specifically to fighting red spider mite (*Tetranychus urticae* Koch) and two spotted spider mite (*Tetranychus cinnabarinus* Bois.).

3) Fungi pathogens of seeds

Again, HEO and hemp extracts proved to be effective in fighting fungi pathogens of seeds in *Apiacae* family and therefore has been patented for this application [29].

Specifically the invention refers to the composition based on extract of sowing hemp used in plant protection against fungi pathogen that affect vigor, germination and health of the *Apiacae* family, especially of carrots.

Occurrence of pathogenic fungi on seeds is one of the reasons behind their poor shooting and the development of the diseases on fields. Therefore the sowing material should be free of pathogens. Seed dressing is the cheapest and most effective method of eliminating pathogens and pests threatening the plants at the early stages of their growth. The most commonly used methods involve using chemical but ecological methods have become more and more popular.

A composition was developed [11,12,18] containing active biologically compounds present in hemp, which are safe for the environment and safer than commonly used chemical substances, that could be used for protection of crop plants.

These applications of HEO in plant protection are only examples of tits potential use. Most likely several new ways of applying will be found as a result of intensive studies of HEO, which are in progress now.

Conclusion

1) The results of long term studies confirmed the possibility of using industrial hemp harvested with a combine harvester as a source of essential oils. The production of HEO is also possible from the panicles of hemp grown for seeds after the threshing.
2) If the main direction of the cultivation of hemp is production of HEO, harvesting panicles at full flowering stage is of key importance.
3) HEO has multiple possible applications in various consumer products.

References

[1] Burczyk H.: *Production of Hemp Sowing Seed in Poland.* Journal of Industrial Hemp, 2003, Vol.8(1), 81-88.

[2] Burczyk H., Grabowska L., Kołodziej J., Strybe M.: *The industrial hemp as a raw material in the energy production.* Journal of Industrial Hemp, 2008, Vol.13(1), 37-48.

[3] Burczyk H., Grabowska L., Strybe M., Konczewicz W.: *Wpływ gęstości siewu i terminu zbioru konopi włóknistych na wydajność biomasy oraz elementów składowych plonu.* Pamiętnik Puławski, Nr 152 Tom 1 2008.

[4] Burczyk H., Strybe M., Konczewicz W., Lukaszewska N., Oleszak G., Baraniecki P.: *Effect of sowing density and time of harvest on seed and panicles yield.* Renewable Resources: Obtaining, Processing and Applying. Editors. F. Pudel, R. Kozłowski, G. Zaikov Publisher: Nova Science Publishers, New York, 2009 ISBN: 978-1-60692-534-8. pp. 35-42.

[5] Burczyk H., Kaniewski R., Konczewicz W., Kryszak N., Turowski J.: *Konopie włókniste źródłem olejków eterycznych.* Pamiętnik Puławski, Nr 151 Tom I, str. 37-47, 2009.

[6] Kaniewski R.: *Steam Distillation of Essential Oils from Hemp Panicles.* Journal of Natural Fibers, volume II, number 1, 2005. New Patents and Techniques.

[7] Kaniewski R.: *Eteryczne olejki konopne.* Biuletyn Informacyjny Polskiej Izby Lnu i Konopi, Nr 6, Poznań, 01.2006.

[8] Kaniewski R.: *Możliwość wykorzystania konopi włóknistych - Cannabis sativa L. w biologicznych preparatach ochrony roślin.* Biuletyn Informacyjny Polskiej Izby Lnu i Konopi, Nr 9, Poznań, 06.2007.

[9] Kaniewski R.: *Instalacja do produkcji eterycznych olejków w zakładzie doświadczalnym IWN w Pętkowie Biuletyn Informacyjny Polskiej Izby Lnu i Konopi,* Nr 10, Poznań, 01.2008.

[10] Kaniewski R.: *Protection composition against pests.* The Belgian and International Trade Fair for Technological Innovation Brussels Innova - Eureka Contest 2009, Brussels, Belgia, 19-21.11.2009.

[11] Kaniewski R.: *Effects of extracts from hemp (cannabis sativa l.) on health, vigour and germination of carrot seeds.* Progress in Plant Protection. Poznań, 2010.

[12] Kaniewski R.: *Health and germination of carrot seeds treated with aqueous and methanol extracts from hemp (Cannabis sativa L.) panicles.* 16th International

Conference on Renewable Resources and Plant Biotechnology NAROSSA®), 7-8.06.2010, Magdeburg, Niemcy.

[13] Kaniewski R.: *Protection composition based on hemp essential oil against pests.* 17th International Conference on Renewable Resources and Plant Biotechnology NAROSSA®), 30-31.05.2011 Poznań.

[14] Kaniewski R., Banach J., D. Kalemba: *The Content of THC in Stems, Leaves, Hemp Panicles and Essential Oil.* 11 th International Conference for Renewable Resources and Plant Biotechnology NAROSSA, Poznań, 06-07.06.2005.

[15] Kaniewski R., Burczyk H.: *New Technology of Harvesting Hemp Grown for Seed.* 14. Journal of Industrial Hemp – production, processing and products. Official Journal of the International Hemp Association. Volume 10, Number 1 2005.

[16] Kaniewski R. Burczyk H.: *New Technology of Harvesting Hemp Grown for Seed.* Journal of Industrial Hemp – production, processing and products. Volume 10, Number 1 2005.

[17] Kaniewski R., Burczyk H., Konczewicz W., Kryszak N., Turowski J.: *Efficiency of hemp essentials oil depending on sowing density and time of inflorescence harvesting.* 15th International Conference on Renewable Resources and Plant Biotechnology NAROSSA®), 8-9.06.2009, Poznań.

[18] Kaniewski R., Dorna H., Jarosz M., Banach J., Szopińska D.: *Zdrowotność i kiełkowanie nasion marchwi traktowanych wyciągiem wodnym z konopi siewnych (Cannabis sativa L.),* H. Dorna, R. Kaniewski, M. Jarosz, J. Banach, D. Szopińska, Progress in Plant Protection, 50(1), 2010, pp. 373-377.

[19] Kaniewski R., Górski R.: *Efficacy of hemp essential oil in the control of rosy apple aphid (Dysaphis plantaginea Pass.) occurring on apple tree.* Progress in Plant Protection 49 (4): 2013-2016. Poznań, 2009.

[20] Kołdowski M., Wysocka-Rumińska A.,Tałałaj S., Wiszniewski J.: *Rośliny Olejkowe i Olejki Naturalne.* PWRiL, Warszawa, 1955.

[21] Konczewicz W., Łukaszewska N.: *"The application of hemp essentials oils in finishing of nonwovens" Renewable Resources: Obtaining, Processing and Applying.* Editors. F. Pudel, R. Kozłowski, G. Zaikov Publisher: Nova Science Publishers, New York, 2009 ISBN: 978-1-60692-534-8. pp. 181-189.

[22] Maier Ch. and Mediavilla V.: *Factors influencing the field and the quality of hemp (Cannabis sativa L.) essential oil.* Journal of the International Hemp Association, 1998, Vol.5(1), 16-20.

[23] Malingre'th H., Hendriks S., Baterman R., Bos J., Visser J.: *The essential oil of Cannabis sativa L.,* Planta Medica, 1975, 28, 56-61.

[24] Me Partland J. M.: *Cannabis as repellent and pesticide.* Journal of the International Hemp Association, 1997, Vol.4(2), 89-94.

[25] Mediavilla V. and Steinemann I.: *Essential oil of Cannabis sativa L. strains,* Journal of the International Hemp Association, 1997, 4(2), 82-84.

[26] Nissen L., Zatta A., Stefanini I. at al., *Characterisiation and antimicrobial activity of essential oils of industrial hemp varietes (Cannabis sativa L.),* Fitoterapia, 81, 2010, pp. 413-419.

[27] Patent submission: *Composition against pests, especially hemiptera, application of the composition for fighting pests and the method of producing the composition.* Submission no. P 389106. Submitted on 23rd September 2009.

[28] Patent submission: *Composition against pests, especially spider mites, the method of producing the composition for fighting pests.* Submission no. P-394918. Submitted on 18.05.2011.

[29] Patent: *Composition based on hemp panicles against pathogens, the method of producing the composition and application of the extract from hemp panicles for producing the composition for fighting pathogens.* Submission no. P 388969. Submitted on 04.09.2009.

In: Organic Chemistry, Biochemistry, Biotechnology ... ISBN: 978-1-62081-155-9
Editors: G. Zaikov, F. Pudel and G. Spychalski © 2013 Nova Science Publishers, Inc.

Chapter VIII

Enriched Linen Textiles as a Material for "Wellness" Products

Romualda Marszałek, Mirosław Wacławek*, Teresa Hernik, Andrzej Sapieja, Barbara Binkowska, Anna Kicińska-Jakubowska and Bożenna Marcinowska

Institute of Natural Fibres and Medicinal Plants, Poznań, Poland

Abstract

Zyrardow. Fabryka Lnu in cooperation with Institute of Natural Fibres and Medicinal Plants branch in Zyrardow have conducted a research concerning linen textiles modified with the use of E vitamin. The purpose of the research was to initiate production of the mentioned textiles with the future use in "wellness" products. "Wellness" is a reflection of a new philosophy of life. In textile industry it refers to functionality of clothing textiles – they have to be comfortable both for body and for soul.

Keywords: Linen fabrics for shirts, dresses, bed sheets, physiological properties, parapharmaceutics, E vitamin, "wellness"

Introduction

Flax fiber is natural cellulosic, fully biodegradable raw material of plant origin known for generations. It is used for yarn production and then fabrics [1]. Nowadays, textiles made of linen fabrics draw attention of the consumers who care about their health. Due to their exceptional features (high hygroscopicity, moisture sorption, and air permeability) linen textiles undergo revival. Being timeless products they symbolize comfort and elegance.

* Linen Factory "Żyrardów. Fabryka Lnu" ul. Karola Dittricha 22, 96-300 Żyrardów, Poland.

However in the face of severe competition on the textile market, it is necessary to introduce new additional functions to linen textiles, which functions are not typically associated with such products. Novel textiles should guarantee comfort and well-being of the user.

This new trend opens a wide spectrum of applications for the 'wellness' textiles with anti-bacterial properties or cosmetic-like activity (caring and moisturizing the skin). The concept of wellness reflects philosophy of lifestyle, which in textile industry translates into functionality of clothing to guarantee comfort of wearing and good feeling. Undoubtedly, clothing enables people to feel better in the hectic life. Such clothing can form unspoken message, about one's personality and at the same time can enhance and complement physiotherapy that reduces surface tension on the skin, which result from stress and fatigue [2,3,4].

With this in mind, Linen Factory in Zyrardow (under project no. 6ZR7 2009C/7155) cooperating with Institute of Natural Fibres and Medicinal Plants in Poznan (Poland) carried out tests on ennobling linen fabrics with a preparation containing E vitamin.

E Vitamin is one of the water-indissoluble and fat-soluble vitamins. It is a strong antioxidant which retards process of aging of the body. From the chemical point of view it is $C_{29}H_{50}O_2$ - compound described as α-Tocopherol.

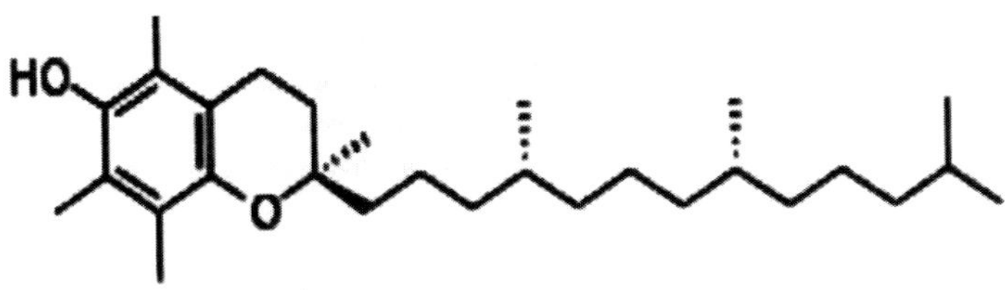

α-tocopherol [2,3]

Figure 1.

This vitamin has been used in the production of pharmaceuticals and cosmetics for years. In cosmetics it serves as antioxidant, soluble in fats and moisture absorber and thus it cares for the skin and enhances treatment of various skin diseases. E vitamin is used in cosmetics containing lipids e.g. creams, ointments, emulsions, face and body oils and lipsticks. Antioxidant is a molecule that neutralizes free radicals, which are metabolites forming during molecular respiration and potentially damaging cell membranes.

To apply E vitamin on textiles it is necessary to use another substance i.e. cyclodextrin (CD), which forms a stable complex with the vitamin. Cyclodextrin is a cyclical molecule containing 6-8 glucose molecules. It has a ring structure with hydrophilic properties outside and due to being non-polar it is hydrophobic or lyophilic. The inside part of the cyclodextrin molecule absorbs non-polar molecules of E vitamin and thus forms a stable complex. Such a complex guarantees higher resistance of the substance to light, oxygen and temperature. The CD-E vitamin complex does not show substantivity to fibers, therefore the application methods include padding, spraying, coating and printing.

Reactive formaldehyde-free polyurethane is used to fix the complex on the fabric, which ensures good durability and softness of resin. Fixing undergoes not only by physical bonding but also chemical, especially in case of cellulosic fibers, where OH-groups are present [2, 3].

Within the discussed project no tests were conducted on the properties of E vitamin itself, as those have been extensively studied in other research centres. The studies on the parameters of the textiles modified with the aforementioned product that influence the physiological comfort of the users.

Aim and Backround

The study aimed at starting a production of innovative linen fabrics for everyday use (shirts, dresses and bed sheets) modified with a natural organic substance (a para-pharmaceutical), which fabrics will improve physiological comfort of users (the so called 'wellness products').

Such fabrics have exceptional properties that determine their market value. The use of flax raw material, characterized with moisture sorption, air permeability and absence of allergenic substances, modified with E vitamin treatment, allowed for developing products with the following features:

- skin care,
- anti-ageing,
- toning and smoothing of skin.

The products with these new „wellness" features are addressed mainly to health-conscious self-confident people. Especially these clothes are recommended for uniforms for office staff of leading companies and for bed sheets for SPA salons and health resorts. These products fulfill requirements for modern textiles and guarantee comfort and well-being of users.

Production of linen fabrics of such features is a novelty on the Polish market considering their fully ecological character. The innovative linen fabrics are competitive towards traditional linen fabrics at the cost of a slightly higher price, not prohibitive for the buyers.

Experimential Part

Starting the production of novel linen fabrics had been preceded by conducting numerous studies at experimental scale, developing of the technological process and complete chemical and metrological analyses of fabrics before and after modification and testing the use parameters of the final products made of these fabrics.

Methods

The presence of E vitamin on the fabrics was determined with colorimetry according to the CHT method (R. Beitlich GmbH, Tübingen) by sprinkling a suitable indicator.

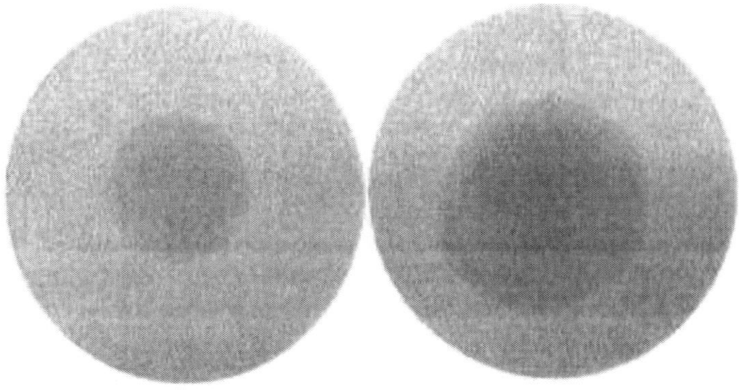

Figure 2.

Sprinkling the $FeCl_3$ solution on the fabric in the presence of E vitamin results in reduction of Fe^{3+} ions to Fe^{2+} ions. Subsequently, by spraying with 2,2'- dipiridil a complex compound of pink colour is formed. The intensity of its colour depends on the content of iron ions (II). This method is applicable only for white fabrics.

During research works the fabrics treated with E vitamin and untreated fabrics were conducted for comparison. The following parameters were tested:

a) Determination of physiological properties – Measurement of thermal and water-vapour resistance under steady state conditions: PN-EN 31092:1998.
b) Determination of permeability of fabrics to air – PN-EN ISO 9237:1998.
c) Determination of hydroscopicity– PN-P-04635:1980.
d) Determination of speed of water absorption in drop method– JIS 1090:1990.
e) Determination of Solar UV protective properties PN-EN 13758-1:2007.
f) Determination of force and elongation using the strip method PN-EN ISO 13934-1:2002.
g) Determination of dimensional change in washing and drying PN-EN 25077:1998, PN-EN ISO 3759:1998.
h) Determination of thread shiting PN-EN ISO 13936-2:2005.
i) Determination of whitening degree PN-P-04775:1988.

1) The evaluated properties:
   a) Aesthetic parameters i.e. the appearance of the products,
   b) Physiological properties i.e. determining the thermal balance and comfort for users,
   c) Mechanical resistance to wear damage,
   d) Ease of care i.e. the method, frequency and efficiency of maintenance procedures that determine the comfort of care.
2) Time of use i.e. minimum time of observing the changes in the product during its use (5 months).

3) Use cycle i.e. the period after which the product is checked for the changes in its characteristics; the term includes the maintenance care (30 hours).
4) Frequency of testing: after 1, 2, 3 and 5 use cycles.
5) Number of users (15 people).

The users were asked to use the products during specified period of time and providing them to INF&MP in Zyrardow for conducting the maintenance procedure and informing the researchers about the behaviour of the products during use and about the users' comfort.

The aesthetic parameters were evaluated by assessment of:

- Shift of threads in the seam,
- Resistance of stains to washing,
- Changes of shape,
- Resistance to pilling.

Determination of physiological properties was based on questionnaires filled in by 15 users. The questionnaire asked about:

- Physiological comfort i.e. a set of properties of clothes that determine users' good feeling by affecting physical and psychological conditions of the user. The assessment is subjective according to a three-grade system: very good, good and unsatisfactory.
- Users' sensations i.e. subjective evaluation based on sensual feeling described usually with the characteristics of the handle: light, thick, thin, fluffy, soft, stiff, supple, and elastic or with expressions like: cold, cool, warm, hot, moist and wet.

The resistance of textile products to mechanical damage is determined according to visual evaluation for abrasions (local absence of one or both thread systems) and tearing (local breaking of one or both thread systems).

There are no standards for linen fabrics refined with E vitamin. Therefore the development of the technology was done in compliance with the existing standards listed below. Adjusting the properties to the requirements of these standards led to developing linen fabrics useful for clothing production that have additional features of enhancing physiological comfort.

- PN-P-82450-02:1986 Linen and linenlike woven fabrics of common use. Performance requirements for shirt woven fabrics.
- PN-P-82450-06:1986 Linen and linenlike woven fabrics of common use. Performance requirements for dress woven fabrics.
- PN-P-82450-4:1996 Textiles. Linen and linenlike woven fabrics of common use - Usably requirements of fillow-case, cover and sheet fabrics.

The targeted parameters of the fabrics treated with pro-healthy ingredients should be as follows (Table 1):

Table 1. Targeted parameters of the refined linen fabrics

Parametr		Unit	Value	Test method
Breaking Force *(shirt woven fabrics)*	warp	daN	≥ 25	PN-EN ISO 13934-1:2002
	weft	daN	≥ 20	
Breaking Force *(dress woven fabrics)*	warp	daN	≥ 18	
	weft	daN	≥ 18	
Breaking Force *(sheet fabrics)*	warp	daN	≥ 45	PN-P-82450-4:1996
	weft	daN	≥ 45	
Thread shiting	warp	mm	≤ 6	PN-EN ISO 13936-2:2005
	weft	mm	≤ 6	
Degree of whitness		%	< 5	PN-P-04775:1988

It was assumed that:

- The textile product is air-permeable if the air-permeability value is higher than 500 mm/s,
- Textiles ensuring good physiological comfort are characterized with water vapour resistance below 10 m^2 Pa/W,
- Textiles characterized with low thermal resistance can be designed for summer clothes. Minimum thermal resistance values for winter clothing protecting from cold are about 0,06 m^2K/W.

Equipment

Industrial equipment of Linen Factory Zyrardow was used for producing the refined fabrics.

Devices used for producing input fabrics in white colour:

- Warping: cone warping machine,
- Weaving: looms STB 175,
- Cropping: Textima,
- Bleaching: Henriksen,
- Drying: dryerTextima.

Devices used for treatment:

- Padding: Fulard 3-roller,
- Drying: dryer Textima and Mascropol TD.

Testing Equipment

- Cary 50 for testing properties of UV protection,
- Air Permeability Tester III for testing air permeability of fabrics,
- SGHP-8.2-Sweating Guarded Hotplate for testing water vapour resistance and thermal resistance,
- Colorimeter d/8 from Spectrocolor for testing whiteness degree of fabrics,
- Tiratest for testing breaking strength and shift of threads in the seams,
- The photos of fabrics improved with E vitamin and without E vitamin have been taken with the use of S-3400N Hitachi Scanning Electron Microscope (SEM) in high vacuum mode (secondary electron detector SE).

Chemicals Applied

For obtaining input white fabrics the following chemicals were used:

- NaOH,
- Washing agent based on polyglycol and ether of fatty alcohols,
- Moisturizing-dispersing agent based on a composition of fatty alcohol ethers and polyglycol,
- Bleaching agent: H_2O_2,
- Stabilizing agent for peroxide baths: water glass.

For the treatment the following agents were used:

- Cross-linking agents, forming in a catalyst free process a soft, highly flexible and formaldehyde free film. Chemical characteristics: micro dispersion of modified polyurethanes,
- Dispersing, washing, moisturizing non-ionic agent. Chemical characteristics: alkyl-polyglycol ether,
- Softening agent that gives the fabric soft handle used for finishing textile products mainly by padding method. Chemical characteristics: functional polysiloxanes in the form of microemulsions,
- Agent containing E vitamin. Chemical characteristics: cyclodextrin complex with E vitamin suspension,
- Acetic acid CH_3COOH.

Results and Discussion

Table 2 presents parameters of linen fabrics treated with agent containing E vitamin and untreated fabrics.

Table 2. Results of tests made on treated (E vitamin) and untreated fabrics

Parameter		Sample			
		30033 biel ¾, treated with agent containing E vitamin	**30033 biel ¾, untreated fabric**	**23080 biel 4/4 treated with agent containing E vitamin**	**23080 biel 4/4 untreated fabric**
Water-vapour resistance [m^2 Pa/W]		4,0653	3,6301	1,8274	1,4456
Thermal resistance [m^2 K/W]		0,0141	0,0109	0,0089	0,0084
Air permeability [mm/s] Right side Left side		 806 833	 519 517	 1543 1546	 1485 1522
Hygroscopicity at 65 % humidity at 100% humidity:		 6,8071 12,8380	 6,8539 14,6585	 7,0060 13,259	 7,1749 13,9569
Speed of water absorption [s]		100,8	76,8	317,4	130,8
Ultraviolet protection factor UPF		15	5	0	0
Presence of E vitamin on the fabric after ten domestic launderings in 60°C		present	-	present	-
Whitening degree [%]		55	55	68	68
Slippage resistance of yarns at a seam in woven fabrics (strength 60 N) [mm]	warp	4	2	3	4
	weft	4	2	3	3
Dimensional change in washing (40°C) and drying [%]	warp	- 1,20	-1,20	- 1,40	-1,40
	weft	- 2,00	-2,00	- 2,00	-2,00
Breaking strength [daN]	warp	60,4	76,2	39,8	45,1
	weft	71,0	93,2	50,7	55,2

The results of the experiment on using the products are listed in Table 3. The presented values are mean calculated from 19 pieces of clothing

Table 3. The results of experimental use of the products made of treated fabrics

No.	Parameters and usable features	Evaluation
1.	Thread shift in seams	No clearances
2.	Crease resistance	During use creases appear but their exposition is limited
3.	Pilling resistance	5 Pilling was not observed
4.	Shrinkage and extensibility of fabric	After 5 maintenance cycles the changes where observed both for shrinkage and extensibility with prevailing shrinkage. The values of these changes do not exceed 1,5%
5.	Change of colour or whiteness degree	During use and maintenance (washing, whirling, drying) not changes were observed
6.	Resistance to mechanical damage/abrasion	No tearing/abrasions observed
7.	Physiological and functional properties	Good air-permeability, proper thermal balance, good feeling and high comfort of using, no restrained movements
8.	Ease of care/maintenance	Gentle washing in a washer, good washability in warm water with commercially available washing agents for colour laundry, softener use recommended, can be whirled, hang drying, ironing damp at 150 °C

The table below shows the SEM images - both fabrics improved with E vitamin and without E vitamin.

The images of fabrics improved with E vitamin and without E vitamin have been taken with the use of S-3400N. Hitachi scanning electron microscope (SEM) in high vacuum mode (secondary electron detector SE). The fabrics, as a non-conductive material, were initially coated with gold in order to ensure proper parameters. The optimal parameters for each fabrics were selected: magnification of 50, stage height of 67mm and accelerating voltage of 23.5 kV were chosen, and for better illustration of the structure of E vitamin application, magnification of 2000, stage height of 15mm and the same accelerating voltage of 23.5 kV were chosen. Thosc papameters allowed for observing the chcaracteristic feattures of the surface of fabrics and elementary fibres [8-11].

Fabrics 23080 improved with E vitamin, magnificatin 50.

Fabrics 23080 without E vitamin magnification 50.

Fabrics 23080 improved with E vitamin, magnification 2000.

Fabrics 30033 improved with E vitamin, magnificatin 50.

Fabrics 30033 without E vitamin magnification 50.

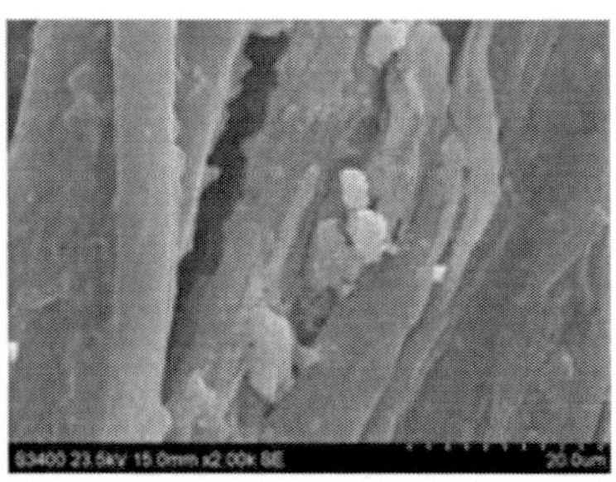

Fabrics 30033 improved with E vitamin, magnification 2000.

Figure 3. SEM images – fabrics improved with E vitamin and without E vitamin.

Discussion

The results presented in Table 2 allowed the following observations.
Regarding the fabric sample no. 30033 whiteness ¾:

- E vitamin was detected at the fabric after the application and after 10 cycles of maintenance, where washing was done at 60°C.
- The refining treatment did not affect the whiteness degree.
- The air permeability for the treated fabric is higher than for the untreated fabric and the value of 806 – 833 [mm/s] indicates that the product is air permeable.
- The UPF value at 15 for the treated fabric shows good protective properties against UV radiation (according to the Australian Standard AS/NZS 4399, 1996). The untreated fabric does not show such properties. UPF reached only 5.
- The treatment contributed to worsening of the thread shift in seams by 2 mm and to lowering of breaking force by 6 daN for warp and by o 22 daN for weft, as compared to the untreated fabric. Despite this, the metrological parameters obtained (thread shift in seams, breaking force, shrinkage and extensibility) allow to apply the treated fabric for shirts, dresses and bed sheets, according to the Polish Standards (Table 1).
- The low value of thermal resistance for the treated fabric (0,0141 m^2K/P) allows for applying it for summer clothing, despite slight increase of this parameter as compared to the untreated fabric (0,0109 m^2K/P).
- The water vapour resistance for the treated fabric (4,0653 m^2Pa/W) is slightly higher than for the untreated one (3,6301 m^2Pa/W). As it is assumed that materials with good physiological comfort are characterized with this value below 10 m^2Pa/W, both fabrics can be regarded as „breathing” fabrics i.e. permeating water vapour.
- For the treated fabric lower hygroscopicity and worse absorption of water drop was observed than respective parameters for the untreated fabrics.

For the fabric sample no. 23080 whiteness 4/4 it was observed that:

- E vitamin was detected at the fabric after the application and after 10 cycles of maintenance, where washing was done at 60°C.
- The refining treatment did not affect the whiteness degree.
- The value of air permeability for the treated fabric at 1500 mm/s indicates that the product is air permeable.
- The value of thermal resistance for both treated and untreated fabrics were at the same very low level what makes the fabrics suitable for summer clothes.
- The water vapour for the treated fabric is slightly higher than for the untreated one. However, its value (1,8274 m^2Pa/W) indicates that the material guarantees very high physiological comfort.
- The zero value of UPF for both treated and untreated fabrics indicates that they do not show UV protection properties.
- The refining treatment, similarly to the case of the sample no. 30033, caused slight worsening of the thread shift in seams by 1 mm and lowering of breaking force by 6

daN, as compared with the untreated fabric. Despite this, the metrological parameters obtained (thread shift in seams, breaking force, shrinkage and extensibility) allow to apply the treated fabric for shirts, dresses and bed sheets, according to the Polish Standards (Table 1).
- For the treated fabric lower hygroscopicity and worse absorption of water drop was observed than respective parameters for the untreated fabrics.

Slight decrease in the water vapour resistance, heat resistance, hygroscopicity and the ability to absorb a water drop in case of refined fabric as compared with untreated fabric does not disqualify the first as a product guaranteeing physiological comfort. The values of physiological parameters of the treated fabrics indicate that the product is air-permeable (> 500 mm/s), with low water vapour resistance (< 10 m^2Pa/W) and low thermal resistance (< 0,06 m^2K/W), what means that it can be used for summer clothing and can ensure physiological comfort.

The values of usable feature parameters (Table 3) allows for making the following observations:

- The users of products with E vitamin gave positive feedback regarding these products. No allergic skin reactions were observed during the use. The products were described as air permeable and the users experience thermal comfort in the clothes.
- Pilling resistance for all tested products was very good reaching the value of 5.
- Resistance to mechanical damage for the tested products was very good; no abrasion and tearing were observed.
- Resistance to thread shift is very good; no clearances were observed.
- The shrinkage after washing varies between 1,5 % and 2,0 % after 5 washing cycles.
- The tested products have good stains to washing with easily available washing agents. The products should be ironed wet at 150ºC.
- All the users noticed that the handle of the fabric changed after each washing i.e. it became less smooth.

Conclusion

- The results of metrological parameters of improved linen textiles indicate that the textile can be used both for production of clothes and bed clothing (according to the PN-P-82450-02, PN-P-82450-06, PN-P-82450-04 requirements - Usably requirements of shirt, dress, sheet woven fabrics).
- After the application of the agent and later conservation process (laundring in 60°C) the E vitamin remains on the textile.
- Parameters' results such as air permeability and thermal resistance indicate that the textile can be used in production of summer clothes.
- Parameters' results such as water-vapour resistance and hygroscopicity combined with high level of airiness and low level of heat-preservation suggest that products made of this textile can provide the user with physiological comfort.

- The value of the UPF indicator (15) shows that the textile preserves well against UV radiation (Australian standard AS/NZS 4399, 1996).

References

[1] G. Urbańczy, Nauka O Włóknie, Wydawnictwo Naukowo-Techniczne Warszawa 1985.
[2] R. Zyschka, R. Brückmann, B. Krammerer, H.Schreiber, Wellness-Finish MIT Vitamin E – Textilveredlung 5/6 2004
[3] W. Michalowski “Wellness” – Wykończenie Tekstyliów Zawierające Witaminę E – Przegląd WOS 11/2005.
[4] W. Michałowski Aktywne Wyroby– Przegląd WOS 7/2006.
[5] Sun Protective Clothing-Evaluation And Classification AS/NZS 4399:1996.
[6] J. Zielińska, S. Kuź, R. Marszałek, Badania Własności Ochronnych Lnianych I Konopnych Wyrobów Włókienniczych Przed Promieniowaniem UV– Przegląd WOS 8/2001.
[7] Techniczne informacje firmy: CHT R. Beitlich GMBH.
[8] A. Kicińska- Jakubowska R. Kozłowski, W. Gieparda Techniques For Nano-Analysis Of Textile And Composites, 5th International Conference of Textile Research Division NRC, Cairo, Egypt, April 6-8, 2008. *Textile Processing: State of the Art & Future Developments* 5(VI) (2008) 306-314.
[9] A. Kicińska-Jakubowska, E. Bogacz, R. Kozłowski, M. Zimniewska,Electron Scanning Microscope Analysis As A More Effective Tool For Studies On Natural Fibres Proceedings of the 4th Global Workshop of the FAO/ESCORENA European Cooperative Research Network on Flax and other Bast Plants. „Innovative Technologies for Comfort”, Arad - Romania, 07-10 October 2007.
[10] Anna Kicińska – Jakubowska, Małgorzata Zimniewska, Ryszard Kozłowski, ANALIZA Porównawcza Wybranych Włókien Naturalnych Przy Zastosowaniu Elektronowego Mikroskopu Skaningowego IX Międzynarodowa Konferencja Bawełniarska “Przyszłość Włókien Celulozowych W Świetle Trendów Rozwoju Przemysłu Tekstylno-Odzieżowego”, Gdańsk 06-07.09.2007.
[11] A.Kicinska – Jakubowska, R. Kozlowski, M. Zimniewska, Scanning Electron Microscope Study Of Natural Fibres IX International Conference on Frontiers of Polymers and Advanced Materials Kraków 8-12 lipca 2007 (Book of Abstracts ISBN 978-83-7204-606-3).

In: Organic Chemistry, Biochemistry, Biotechnology ... ISBN: 978-1-62081-155-9
Editors: G. Zaikov, F. Pudel and G. Spychalski © 2013 Nova Science Publishers, Inc.

Chapter IX

Polish Cultivar of White Mulberry (*Morus Alba* L.) and New Directions of Its Utilization

***Małgorzata Łochyńska*[1,*], *Grzegorz Oleszak*[2] *and Natalia Kryszak*[3]**
[1]Department of Silkworms Breeding and Mulberry Cultivation, Institute of Natural Fibres and Medicinal Plants, Poznan, Poland
[2]Experimantal Farm INFandMP in Petkowo, Środa Wielkopolska
[3]Department of Harvesting Technology and Bast Plants Evaluation, Institute of Natural Fibres and Medicinal Plants, Poznan, Poland

Abstract

Polish cultivar of white mulberry Żółwińska was bred in 1950's in Milanówek, near Warsaw, Poland. It characterizes with huge leaves and rapid growth, because this cultivar was designed for mulberry silkworms (*Bombyx mori* L.) breeding. These days, when the European sericulture undergoes crisis, sericulturists must find new directions of mulberry utilization. The aim of this paper is to show the studies on the Polish cultivar of white mulberry and propose future course for the development of the industry.

Keywords: White mulberry, utilization, energy industry, food industry, fatty acids

[*] E-mail: malgorzata.lochynska@iwnirz.pl.

Introduction

Morus alba Linnaeus, 1753 is one of the numerous species in the family Moraceae. Most of the species are native to Asia. These plants are characterized with milky sap in shoots. They are both monoecious and dioecious, their flowers are inconspicuous and odorless and the small, sweet fruits are multiple: drupelets or nuts. The leaves are alternately arranged, simple, often lobed and serrated on the margin (Litwińczuk 1993; Butt 2008).

There are numerous uses of white mulberry known for centuries. It was well known that infusion and tea of leaves are healthy and they were considered as diaphoretic and emalliorent and applied for gargling in inflammations of throat. The fruits give cooling effect and are used as a laxative, the roots possess anthelmintic activity and astringent properties and the bark is used as a purgative and vermifuge (Sharma 1994). However, white mulberry was mainly known for its excellent antidiabetic action.

These days we know, that white mulberry contains a lot of active compounds, which are very precious for human health. The leaves and fruits contain 15-31% high quality protein, crude fiber, 200-300 mg/100g of ascorbic acid, of which over 90% is present in the reduced form, vitamin B, D, folic acid, folinic acid, β-carotenes and essential macroelements (K, Ca, Mg, Na, P) and microelements (Fe, Zn, Ni) (Srivastava 2006; Ercisli 2007; Imran 2010). Moreover, there is a lot of flavonoids (quercetin, rutin, isoquercitrin, astragalin, kuwanon G and C, catechin, mulberrofuran G, albanol B, morusin, sanggenon B and D), which exhibit strong antioxidant activity (Kim 1999; Doi 2001, Kofujita et al. 2004), antifungal and antibacterial potential against harmful strains of bacteria (Shirata 1982; Park 2003) and against viruses HSV-1 (Kimura 2007; Butt 2008). Moreover, mulberry flavonoids lower significantly blood glucose levels by inhibiting enzyme activity (Oku 2006), prevent and inhibit of atherosclerosis by strong inhibitory effects on LDL oxidation and increasing resistance to blood cholesterol deposits (Chen 2006, Butt 2008). What is more, the polysaccharides from the bark of mulberry roots stimulate lymphocyte proliferation and reduce the production of antibodies (Kim 2000, Butt 2008). In addition, the mulberry cyanidin protects the brain against endothelial dysfunction and reduces the likelihood of Alzheimer's disease (Serraino 2003). The literature data also reported considerable capacity of mulberry for local whitening of skin, so called depigmentation (Fang 2005).

There are other ways of using this multipurpose tree. Leaves of white mulberry were and still are used as a fodder for mulberry silkworms (*Bombyx mori* L.). In India all annual shoots, available after silkworm breeding season, are cut, dried and used as the main renewable source of fuel (Chinnaswamy 1995). Moreover, mulberry may be used as a natural dyeing plant (Sharma 1994), in the paper, furniture and sport equipment industry and as a substrate for oyster mushrooms *Pleurotus* sp. (Madan 1992). Mulberry leaves (not only fruits) are eaten by vegetarians, both leaves and fruits after drying are applied as fodder for birds, reptiles and rodents. In addition, mulberry may be also used in gardening.

The aim of this paper is to present the old, Polish cultivar of white mulberry Żółwińska and find new directions of mulberry use in Polish conditions.

Material and Methods

All studies are carried out on an old, Polish cultivar Żółwińska, which was bred in 1950's in Milanówek, near Warsaw, Poland. These days there is established a proecological plantation in Experimental Farm INFandMP in Petkowo (52,12N; 17,15E). The studied material was: shoots, leaves and fruits of Żółwińska cultivar and leaves and shoots of China and Kokuso cultivars (from experimental plantation in Petkowo). There were measured: morphological features, energy value of shoots, cellulose and pectin contents in shoots and fatty acids content in seeds.

The morphological features were carried out in July 2010. There were measured: length and width of the leaf, petiole length, distance from the leaves on the shoot, leaf color, fruit length and the stalk of a fruit. One hundred samples for each examination were taken. Present measurements of Żółwińska cultivar were compared with two the most similar cultivars: Kokuso and China (with exception of fruits studies). All mulberries are cultivated in the same plantation conditions.

The energy value of white mulberry Żółwińska was examined at the Department of Environment Protection INFandMP. The dried annual shoots collected in July 2009 were used in the investigation. The research was carried out in the oxygen bomb calorimeter type KL-5.

Content of cellulose and pectin was examined in the Harvesting Technology and Fibrous Raw Material Evaluation Department INFandMP. The material – annual shoots – was collected in June 2010, cut into 3-4 cm peaces and died. Samples were melt in the mill Pulverisette 19 and then wax and grease substances were removed. After chemical treatment the cellulose content was calculated according to the pattern:

$$C = \frac{m_1 x(100 + W)}{m_0} \ (\%)$$

where m_l – constant mass of cellulose (in g), W – moisture of sample (in %), m_0 – initial mass.

After chemical treatment the content of pectin was calculated according to the pattern:

$$P = \frac{m_1 x5}{m_0} \text{ x } 100 \ (\%)$$

where m_l- constant mass of pectin (in g), m_0 – dry sample weight (in g), 5 – factor.

Fatty acid content was studied in the Department of Linseed and Hemp Seed Research and Processing INFandMP. Seeds were collected from fresh black fruits, dried and melt. Extract of seeds after chemical treatment was examined in the gas chromatograph Perkin Elmer with flame detector (FID). Computer software TurboChrom makes all calculations of the fatty acid profile according to the pattern:

Pk [%] = P1/(Pc)

where Pk – % fatty acids, P1 – % P1 surface of detected fatty acid, Pc – [%] (P1+ P2+P3...Pn).

Results and Discussion

The morphological measurements show that the Polish cultivar Żółwińska possesses the best features within compared cultivars. The leaves are the biggest, the petioles are the longest, shiny leaves are dark green coloured, with deeply serrated margins, the average length of black fruits is 1,59 cm and they are set on stalk (average length 0,8 cm). Such a huge leaves possess higher content of precious active substances (eg. flavonoids, vitamins and minerals) what is valuable for medicine and pharmaceutical industry. Moreover, bigger leaves are much better food for mulberry silkworms. Present studies do not include fruit measurements of China and Kokuso cultivars, because these mulberries are too young to fruit. In the future there are planned more detailed measurements. The comparison of studied cultivars features is given in Table 1.

The cellulose content of Polish cultivar was calculated on level 30,27 % for annual shoots (moisture 8,38 %) and 33,59 % for two years old shoots (moisture 7,65 %). The results were similar for China and Kokuso mulberries (Figure 1).

Table 1. The comparison of morphological features of studied cultivars of white mulberry (all measurements in cm)

Feature	China cultivar	Kokuso cultivar	Żółwińska cultivar
length of leaf	10,55	10,45	19,45
width of leaf	7,7	8,25	12,34
length of petiole	2,38	2,06	4,49
distance between petioles	6	4,64	5,8
color of leaf	light green, mat	light green, mat	dark green, shiny

Present studies give worse results than that of previous mulberry examinations (Sharma 1994), where were noted much higher results: cellulose (57,4%), hemicellulose (16,3%) and lignin (24,6%). This difference may be due to features of another cultivar or different method of analyses, which was not given.

The pectin content of Żółwińska cultivar was calculated on level 6,14 % for annual shoots and 3,34 % for 2 years old shoots. This decreasing trend was recorded also in other studied mulberries (Figure 2).

The huge energy value – 17,9 MJ/kg was recorded. This value is a little bit higher than earlier noted (17,0 MJ/kg in Li et al. 2009). Moreover, it is possible to obtain 14-17 tones/ha of biomass (Sharma 1994; Łochyńska 2011). This energy potential, considerable resistance to disease and pests and relatively low soil requirements make the white mulberry a valuable plant for energy industry. Mulberry plantations provide a low utilization of productive or degraded agricultural land. Therefore, it may be used to produce heat, electricity, in fuel production and as slurry in anaerobic digesters.

The fatty acids examination shows that within 13 recorded acids, 7 are mono- and polyunsaturated fatty acids (MUFAs and PUFAs), which belong to very precious fatty acids group: omega-3. The highest content show: linoleic acid (LA, C18:2n6c; 76,84%) and oleic acid (C18:1n9c; 7,09%) (Figure 3).

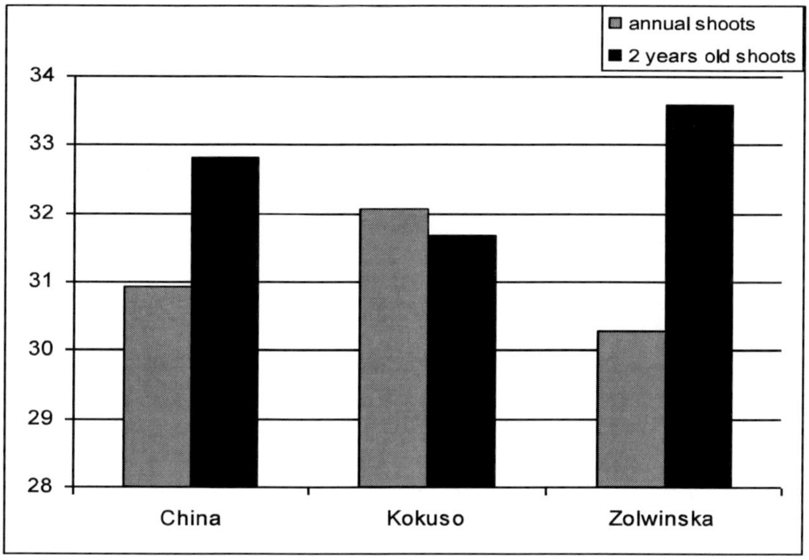

Figure 1. Comparison of cellulose content of China, Kokuso and Żółwińska cultivars (in %).

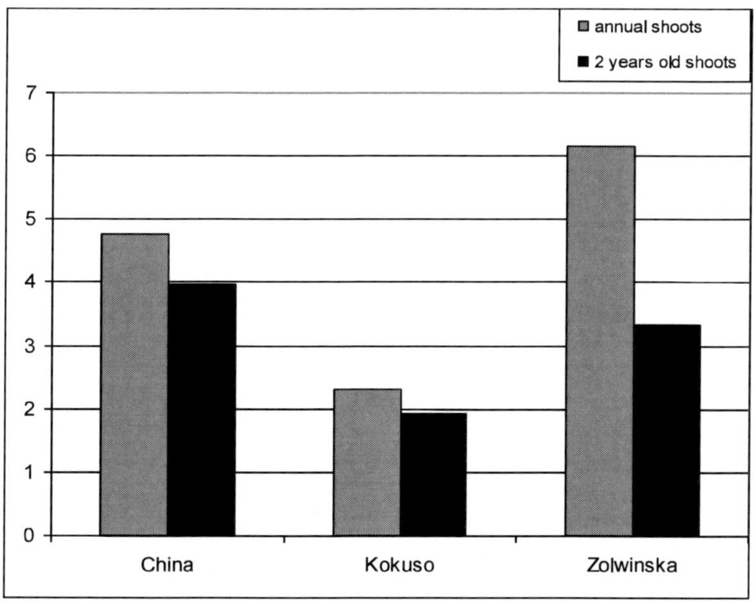

Figure 2. Comparison of pectin content of China, Kokuso and Żółwińska cultivars (in %).

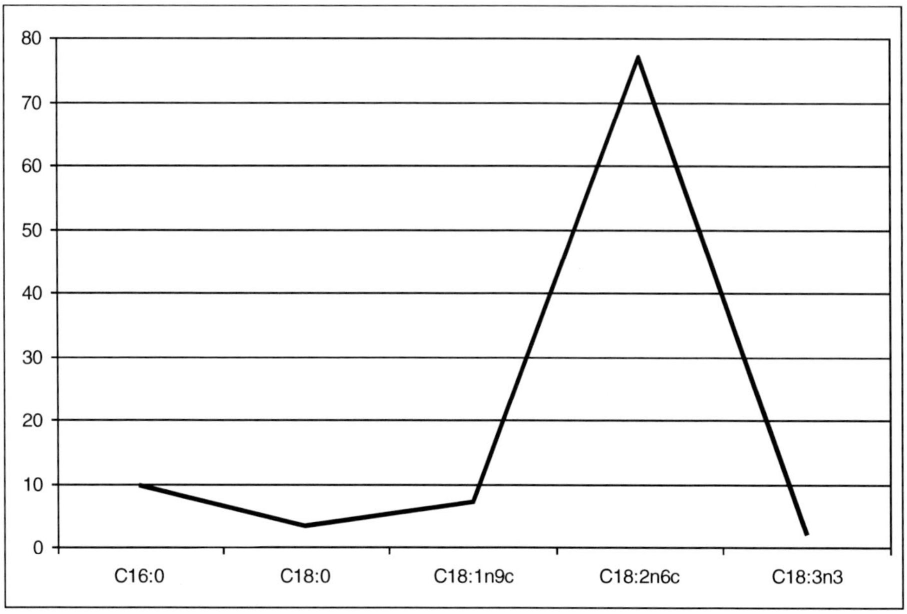

Figure 3. Selected fatty acids content of Żółwińska cultivar (in %).

Trace amounts of other unsaturated fatty acids: linolenic, stearidonic, eicosadienoic, eicosanoic and erucic acid were also recorded. In consideration of this, that the mulberry seeds contain 25-35% of a yellow oil, white mulberry may used as an oil-bearing plant. Detected unsaturated fatty acids are essential unsaturated fatty acids (EUFA), so called vitamin F, which cannot be produced in human organism and must be supplied with food. Fatty acids omega-3 are very precious for our health. They prevent arteriosclerosis, thrombosis and embolism, so that they are used in prevention of heart disease. They support the development and normal brain function and visual acuity, reduce level of LDL cholesterol in the blood and the risk of cancer. Moreover, fatty acids omega-3 strengthen our body by: weakening of the inflammatory symptoms, optimising development of the nervous system, assisting the process of concentration and reduced risk of depression.

Conclusion

Undoubtedly, this multipurpose plant may be used not only in sericology. These days, when white mulberry is used also in sport equipment industry, in furniture industry, as a dyeing plant, in paper industry, in production of oyster mushrooms, as organic fertilizer, as binding and cooperage materials, as fodder for animals, it is believed that huge potential of white mulberry still is not exhausted. There is a huge need to use this precious plant in pharmaceutical industry, medicine and food industry in bigger scale in the age of lifestyle diseases (diabetes, atherosclerosis, hypertension). Moreover, still increasing demand for energy causes the pressure for new sources of renewable energy, what gives a large chance to white mulberry. It is hoped that this very valuable plant will be appreciated not only in Poland but also in other countries of EU, because the use of mulberry potential certainly will develop several economic sectors.

References

Butt M.S., Nazir A., Sultan M.T. and Schoën K., 2008, *Morus alba* L. nature's functional tonic, *Trends in Food Science and Technology* 19: 505-512.

Chen P.N., Chu S.C., Chiou H.L., Kuo W.H., Chiang C.L. and Hsieh Y.S., 2006, Mulberry anthocyanins cyanidin 3-rutinoside and cyaniding 3-glucoside exhibited an inhibitory effect on the migration and invasion of a human lung cancer cell line, *Cancer Letter* 235: 248-259.

Chinnaswamy K.P. and Hariprasad K.B., 1995, Fuel Energy Potentiality of Mulberry, *Indian Silk* 34 (4): 15-18.

Doi K., Kojami T., Makino M., Kiura Y., Fujimoto Y. 2001. Studies on the constituens of the leaves of *Morus alba* L, *Chemical and Pharmacology Bulletin* 49: 151-153.

Ercisli S., Orhan E. 2007. Chemical composition of white (*Morus alba*), red (*Morus rubra*) and black (*Morus nigra*) mulberry fruits. *Food Chem.* 103: 1380-1384.

Fang S.H., Hou Y.C., Chao P.D. 2005. Pharmacokinetic and pharmacodynamic interactions of morin and cyclosporine. *Toxicology and Applied Pharmacology* 205: 65-70.

Imran M., Khan H., Shah M., Khan R., Khan F. 2010. Chemical composition and antioxidant activity of certain *Morus* species. *Biomed. and Biotechnol.* 11(12): 973-980.

Kim S.Y., Gao J.J., Lee W.C., Ryu K.S., Lee R.R., Kim Y.C. 1999. Antioxidative flavonoids from the leaves of *Morus alba*, *Archiv der Pharmazie* 22: 81-85.

Kim H.M., Han S.B., Lee K.H., Lee C.W., Kim C. Y., Lee E.J. 2000. Immunomodulating activity of a polysaccharide isolated from Mori Cortex Radicis. *Archives of Pharmacological Research* 23: 240-242.

Kimura T., Nakagawa K., Kubota H., Kojami Y., Goto Y. and Yamagishi K., 2007, Food-grade mulberry powder enriched with 1-deoxynojirimycin suppresses the elevation of postprandial blood glucose in humans, *Journal of Agricultural and Food Chemistry* 55: 5869-5874.

Kofujita H., Yaguchi M., Doi N., Suzuki K.. 2004. A novel cytotoxic prenylated flavonoid from the root of *Morus alba*, *Journal of Insect Biotechnology and Sericology* 73: 113-116.

Li L., Ya T., Jia-sui X., Yuan-liang Y. 2009. The role of marginal agricultural land-based mulberry planting in biomass energy production. *Renewable Energy* 34: 1789-1794.

Litwińczuk W. 1993. Characteristic, propagation and use of white mulberry (*Morus alba* L.). *Biuletyn Ogrodów Botanicznych* 2: 27-35 (in Polish).

Łochyńska M., Oleszak G. 2011. Multi-use of the white mulberry (*Morus alba* L.). *Ecological Questions*, in print.

Madan M., Sharma S. and Vasudevan P., 1992, Mineral content of *Pleurotus sajor-caju* and or- ganic substrates used, *Microbios* 69 (279): 113-118.

Oku T., Hamada M., Nakamura M., Sadamori N. and Nakamura S., 2006, Inhibitory effects of extractives from leaves of *Morus alba* on human and rat small intestinal disaccharidase activity, *British Journal of Nutrition* 95: 933-938.

Park K.M., You J.S., Lee H.Y., Baek N.I. and Hwang J.K., 2003, Kuwanon G: an antibacterial agent from the root bark of *Morus alba* against oral pathogens, *Journal of Ethnopharmacology* 84: 181-185.

Serraino I., Dugo L., Dugo P., Mondello L., Mazzon E. and Dugo G., 2003, Protective effects of cyanidin-3-Oglucoside from blackberry extract against peroxynitrite-induced endothelial dysfunction and vascular failure, *Life Science* 73: 1097-1114.

Sharma S., Madan M. 1994. Potential of mulberry (*Morus alba*) biomass. *Journal of Scientific and Industrial Research* 53: 710-714.

Shirata A. 1982. Antifungal activity of bark pieces of mulberry (*Morus alba*) cultivar Ichinose shoot and factors causing its change. *Annals of the Phytopathological Society of Japan* 48: 147-52.

Srivastava S., Kapoor R., Thathola A. and Srivastava R.P, 2006, Nutritional quality of leaves of some genotypem of mulberry (*Morus alba*), *International Journal of Food Science and Nutrition* 57: 305-313.

In: Organic Chemistry, Biochemistry, Biotechnology ... ISBN: 978-1-62081-155-9
Editors: G. Zaikov, F. Pudel and G. Spychalski © 2013 Nova Science Publishers, Inc.

Chapter X

Evaluation of Different Metal Oxide Nanoparticle Finishes for Self-Cleaning of Cotton Fabric

Sheikh Muhammad Nawaz*[*]*, Babar Shahbaz, Zeeshan Nawaz and Rameez Ibraheem
Department of Fibre Technology, Institute of Microbiology,
University of Agriculture, Faisalabad, Pakistan

Abstract

Nano coating the surface of textile is one approach to the production of high active surface to have ultra violet blocking, antimicrobial and self-cleaning properties. Self-cleaning effect plays an important role in textiles, especially for military protection wears and hospital use. Metal oxide nano particles resist against stains and remove it by self-cleaning process. Nano coating the surface of textiles is an advance towards the production of highly active surfaces to have antimicrobial, self-cleaning as well as ultra violet blocking properties.

Keywords: Nano coating, Self cleaning, Ultra violet blocking, Special metal oxides and Antimicrobial properties

[*] Address of corresponding author: Professor Sheikh Muhammad Nawaz - Department of Fibre Technology, University of Agriculture, 931-B People's Colony, City: Faisalabad, Country: Pakistan. Phone: 0092 41 8549169. Cell: 0092 345 7783160. E-mail: professormnawaz@yahoo.com.

Introduction

A special class of metal oxide nano size particles is used for imparting self-cleaning as well as anti-bacterial properties. These metal ions and metallic compounds exhibit a certain degree of sterilizing effect. A part of the oxygen in the air and water is converted into active oxygen through a catalyst containing the metallic ion, thereby destroying the organic substance to create a sterilizing effect. Nano particles possess enhanced catalytic abilities due to their highly stressed surface atoms which are very reactive. In the use of nano sized particles, the number of particles per unit area is enormously increased [1]. Nano-Ag has been used for imparting antibacterial properties and zinc oxide nano-particles for antibacterial and ultra violet blocking properties. Zinc oxide and titanium dioxide are non-toxic and chemically stable under exposure to both high temperatures and ultra violet. Furthermore, nano-particles have a large surface area to volume ratio that results in a significant increasing of the effectiveness in blocking ultra violet radiation when compared to bulk materials [2]. In this part, it will be relevant to discuss about titania and functional mechanisms. Titanium dioxide is one of the most popular and promising materials in photo catalytic application due to its strong oxidizing power of its holes, high photo-stability. UV tests indicate a significant increment of the UV absorbing activity in the TiO2-treated fabrics. Such result can be exploited for the protection of the body against solar radiation and for other technological applications. Thus, it could be seen that the 100%cotton and polyester/cotton fabrics treated with nano TiO2 had various functionalities, such as antimicrobial activity, self-cleaning and UV protection functions [3]. Photo catalytic activity of TiO_2 and MgO nano particles can break harmful and toxic chemicals and biological agents. These nano particles can be pre-engineered to adhere to textile substrates by using spray coating or electrostatic methods. Finishing with nano particles can convert fabrics into sensor based materials[4]. Self-cleaning fabrics containing TiO_2 nano particles have been developed based on their photo catalytic ability of oxidizing dirt and other contaminants. TiO_2 treated materials showed self-cleaning activity under daylight environment. Microbial protective clothing has been prepared by coating military garments with silver nano particles. These coatings in addition make the fabric self-wound healing, water-repellent, and dirt-repellent. Research is also going on to prepare the chemical and biological protective clothing with self-decontamination and drug-delivery characters. The decontaminants like nano metal oxides (MgO, Al_2O_3) along with activated carbon and antibiotics in the form of nano coatings is being explored for this purpose [5]. The organic pollutants decompose to simple inorganic compound such as CO_2, H_2O and various mineral acids, which reenter the atmosphere and wash away under the effect of heat, wind or rain, so that the coating are self-cleaning with an efficiency which is dependent upon the degree of photo catalytic activity of the MO_2 catalyst, and directly proportional to the total surface area of the MO_2 particles to which the pollutant are exposed [6]. Titanium coating on fabrics acts as an anti bacterial photo catalyst that helps to break down carbon-based molecules. Once triggered by sunlight, clothing made of such fabrics will be able to remove dirt, pollutants and micro-organism by itself [7].

Review reveals that mostly TiO_2 and ZnO nano metal salts are known to impart self-cleaning attribute. On the other hand in the present researches other salts of similar color and features viz; SiO_2 and AL_2 O_3 are clubbed in order to explore their self-cleaning aptitude and other functionalities.

Materials and Methods

The research work upon different metal oxide nano particle finishes for self-cleaning of cotton fabric" was initiated in the Department of Fibre Technology, University of Agriculture, Faisalabad and mainly conducted at Bismillah Textile Mills Ltd. Faisalabad, Pakistan. The cotton fabric processing details, and methods applied to record the effect of different self-cleaning finishes on cotton fabric processing and performance are described here under.

Finish Type	Concentration (Nano particle wt% on weight of fabric)	Particle Size (nm)
$F_1 = TiO_2$ $F_2 = ZnO$ $F_3 = SiO_2$ $F_4 = AL_2 O_3$	$C_1 = 0.5$ $C_2 = 1.0$ $C_3 = 1.5$ $C_4 = 2.0$ $C_5 = 2.5$	$S_1 = 10 \pm 2$ $S_2 = 20 \pm 2$ $S_3 = 30 \pm 2$

This study was carried out step-wise as under:

- Testing of grey fabric as per standard techniques of ASTM (2010).
- Singeing and Desizing
- Scouring and Bleaching
- Dyeing of bleached fabric samples
- Finishing (Curing)
- Analysis of self cleaning action and physical characteristics of fabric as per standard techniques of ASTM and AATCC [14].

Fabric Quality Evaluation

All fabric quality parameters were tested under standard atmospheric conditions (Temperature = $20 \pm 2\ ^0$C and relative humidity $65 \pm 2\%$).

Ultra Violet Blocking Test

Cotton fabric specimens treated with metal oxide nano particle finishes were placed in sunlight for 24-48 hours. During first 24 hours exposure, samples absorbed 35% UV light while during next 24 hours they absorbed 75% UV light. The increased surface area due to the coating of metal oxide nano particles finishes and their uniform distribution on the fabric surface increased the UV absorption efficiency. Ultraviolet measurement was performed as per the technique adopted in a similar study [13].

Evaluation of Self-Cleaning Action

Multiple stains were applied on (2"x2") cotton fabric samples and allowed to spread. One half of each stain on the fabric was exposed to sun light for 12-48 hours while the other half was covered with a block paper to prevent its irradiation from sunlight. The exposed part of the stain was compared with that of the covered part for self cleaning action.

The rate of degradation of stain was faster during the first 12 hours of exposure (almost 30-50% of change was observed in initial 12 hours). Test procedure was adopted from an identical research work [13].

Analysis of Data

The data thus obtained was statistically analyzed applying analysis of variance technique using factorial experiment with replications[15] via statistical packages for social science SPSS version 10.0 [16].

Results and Discussion

The data pertaining to the comparative research study on the effect of different metal oxide nano particle finishes for self-cleaning prospective of cotton fabric was subjected to statistical manipulation and the findings are presented in tables. The outcome is discussed here under character-wise along with their statistical analysis.

Ultra Violet Blocking Test

Statistical analysis of data regarding fabric UV blocking test under the effect of four different nano particle self-cleaning chemicals, five different concentration of binder to chemicals, three nano sizes presented in table 1, reveals non-significant effect of nano-particle finishes (F), highly significant effect for concentrations (C) and nano-particle sizes (S) on metal oxide nano particle coated cotton fabric UV blocking capabilities.

The individual comparison of mean values of coated fabric UV blocking potential under four types of finishes presented in Figure (1) elaborates non-significant differences among mean values. The best performance is attained by ZnO as 70.91%, while the respective mean value of F1, F3 and F4 are 70.84, 70.84 and 70.85 percent. It is inferred that the UV blocking of the fabric is greater in case of ZnO coating. Which exhibits comparatively greater potential to absorbs UV light than rest of the inline finishes. In control fabric, an average of 20% of UV light was absorbed, while the rest passed through the fabric. In case of bulk-ZnO coated fabric, about 50% of the UV light was absorbed by the fabric. In case of nano-ZnO coated fabric, a maximum of 75% absorption of UV light was noticed. The increased surface area of nanoparticles and their uniform distribution on the fabric surface might have increased the UV absorption efficiency [8]. Nano-Ag has been used for imparting antibacterial properties and ZnO nano-particles for antibacterial and UV-blocking properties. Zinc oxide and titanium dioxide are non-toxic and chemically stable substances, under exposure to both high temperatures and UV radiations. Nano-particles have a large surface area to volume ratio that results in a significant increasing of the effectiveness in blocking UV radiation when compared to bulk materials [2]. After water treatment at 100 °C or higher, the cotton fabric was covered with approximately 24 nm diameter needle-shaped ZnO nano rod and had an excellent UV-blocking property with UV protection factor value over 50 [9].

The individual comparison of mean values of treated fabric sample, UV blocking under five types of concentrations, presented in Figure (1) elaborates highly significant differences. The best performance is obtained for C5 as 72.56%, while the respective mean values for C4, C3, C2 and C1 are as 72.02 %, 71.03 %, 69.66 % and 69.04 %. These results indicated that as the concentration of finish and binder increases, the absorbency of UV light increases. The concentration C5 absorbed more UV light than other concentrations, no doubt rest of the concentrations showed equally good absorbency for UV light. The undyed and untreated fabrics (woven or knitted) have high UV transmission and, therefore, little protection. However, when dyed, the dyed fabrics could provide better protection than undyed fabrics. This protection depends on the type of dye, its concentration and the type of fabrics. The chemical cross-linking of undyed fabric with UV-absorber appears to be the best way to increase its UPF and to provide an excellent UV protection [10]. Clay nano particles possess electrical, heat and chemical resistance and an ability of blocking UV light. Therefore, composite fibres reinforced with clay nano particles exhibit flame retardant, anti-UV and anti corrosive behaviors [4].

The individual comparison of mean values of treated fabric UV blocking for three different particle sizes presented in table (1a); elaborates highly significant differences. The best value is obtained for S1 as 73.62 %, while the respective mean value of S2 and S3 are as 71.06 and 67.91 %. These results indicate that particle size S1 possesses greater UV light absorbency, which reflect that as the particle size decreases the absorbency factor for UV light increases on the other hand when the particle size increases, the absorbency of UV light decreases. UV tests indicate a significant increment of the UV absorbing activity in the TiO2-treated fabrics. Such result can be exploited for the protection of the body against solar radiation and for other technological applications. Thus, it could be seen that the 100%cotton and polyester/cotton fabrics treated with nano TiO2 had various functionalities, such as antimicrobial activity, self-cleaning and UV protection functions [3]. The cotton fabric was padded with the nanosol solution, dried at 60°C, and cured at 150°C. Scanning electron microscopy showed continuous and uniform film on the fiber surface. Excellent UVR scattering was obtained with all treated fabrics. Increasing titania content in the nanosol solution leads to increased UVR protection. The UV protection of textile fabrics depends on the construction parameters. As higher fabric density results in a higher fabric surface mass and a higher fabric cover factor, UV-A and UV-B transmission decreases and UV protection increases. Woven fabrics shrink during alkali scouring. Which result in increasing fabric density and cover factor and UV protection. Plain woven cotton fabric due to the higher filling factor, not only the fabric type of weave but also the warp and weft density, yarn structure and fineness, have an important role in UV protection [10].

Table 1. Analysis of variance for UV Radiation Blocking (%)

S.O.V	D.F	S.S	M.S	F.Value	Prob.
F	3	0.038	0.013	2.35	N.S
C	4	108.261	27.065	4985.67	0.0000**
S	2	327.005	163.502	30118.6	0.0000**
FxC	12	0.065	0.005	1.00	N.S
FxS	6	0.039	0.006	1.18	N.S
CxS	8	0.589	0.017	1.21	N.S
Error	24	0.130	0.005		
Total	59	458.236			C.V = 0.1%

* = Significant Ns = Non-significant ** = Highly Significant.

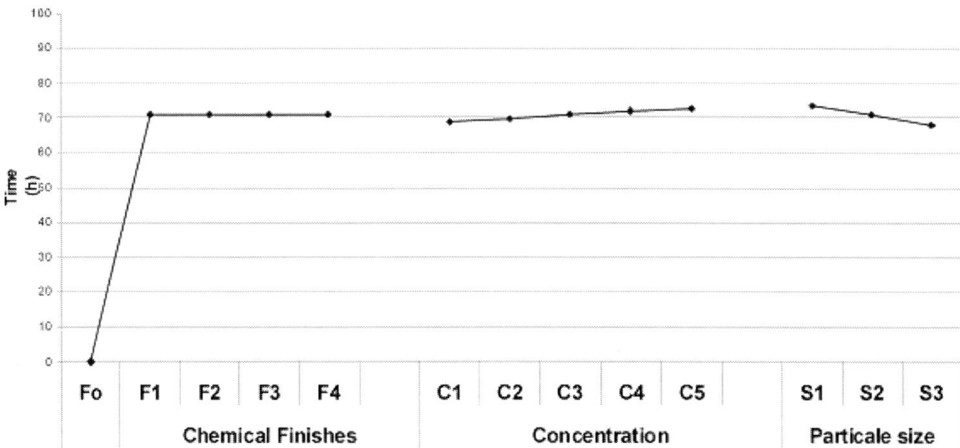

Figure 1. Graphical representation of mean values of Fabric for UV-Blocking (%).

Nano Coated Fabric Self-Cleaning Time to Roll off Ink Stain (H)

The statistical analysis of data regarding treated fabric self-cleaning attribute under the effect of four different nano particle self-cleaning chemicals, five different concentration of binder to chemicals, thre e nano particle sizes presented in table 2, reveals highly significant effect for nano-particle finishes (F), concentration (C) as well as nano-particle size (S) on cotton fabric self cleaning capabilities.

The individual comparison of mean values of treated fabric under the action of four types of finishes upon self-cleaning efficacy presented in table (2a) elaborates highly significant differences. The best value is obtained under TiO_2 finish as 41.09 h, while the respective mean values for F2, F4 and F3 are as 41.94, 42.61 and 42.71 h. It is inferred that when stain of ink is applied on the fabric, the chemical finish F1 (TiO_2) come first with minimum time for the self-cleaning of fabric. When ink is applied for staining on fabric and the specimen is exposed to sun light for self-cleaning action. The initial action started after 6 hours exposure and with in 41 hours the process of self-cleaning is complete. For the self-cleaning of fabric the chemical nano finish ZnO is the second best finish after F1. Photo catalytic activity of TiO_2 and MgO nano particles can break harmful and toxic chemicals and biological agents. These nano particles can be pre-engineered to adhere to textile substrates by using spray coating or electrostatic methods. Finishing with nano particles can convert fabrics into sensor based materials [4]. The titanium coating on fabrics acts as an anti bacterial photo catalyst that helps to break down carbon-based molecules. Once triggered by sunlight, clothing made of such fabrics will be able to remove dirt, pollutants and micro-organism by itself [7]. Actually the organic pollutants decomposes to simple inorganic compound such as CO_2, H_2O and various mineral acids, which reenter the atmosphere and wash away under the effect of heat, wind or rain, so that the coating are self-cleaning with an efficiency which is dependent upon the degree of photo catalytic activity of the MO_2 catalyst, which is directly proportional to the total surface area of the MO_2 particles to which the pollutant are exposed [6]. There is substantial increase in self cleaning or stain release property of the nano TiO_2 treated fabrics. The self cleaning effect obtained by the nano-surface modification is independent of the nature of the substrate. It has been seen that the intensity of self cleaning changes due to the changes in the size characteristics of the TiO_2 nano-particles [3].

Table 2. Analysis of Variance for ink stain self-cleaning time of fabric (H)

S.O.V	D.F	S.S	M.S	F.Value	Prob.
F	3	25.119	8.373	51.82	0.000**
C	4	21.671	5.417	33.53	0.000**
S	2	26.913	13.456	83.29	0.000**
FxC	12	2.801	0.233	1.44	N.S
FxS	6	0.987	0.164	1.02	N.S
CxS	8	1.543	0.192	1.19	N.S
Error	24	3.877	0.161		
Total	59	82.913			C.V= 0.96 %

* = Significant, Ns = Non-significant, ** = Highly Significant.

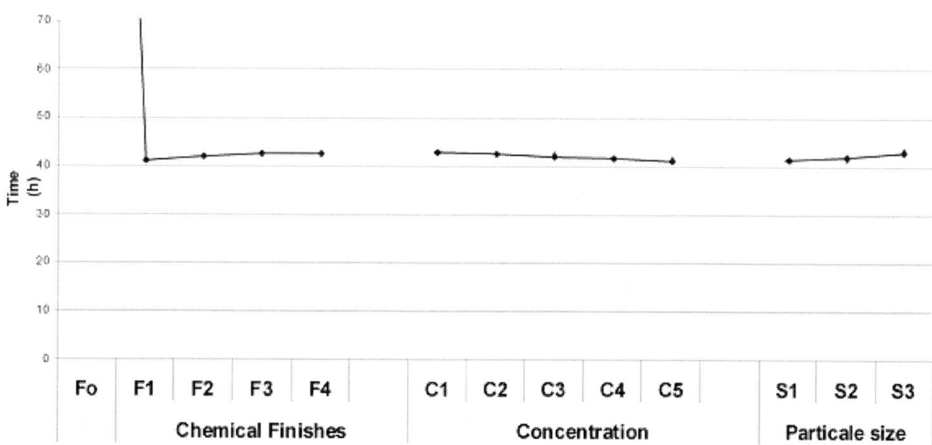

Figure 2. Graphical representation of Fabrics for self-cleaning of ink stains (H).

The individual comparison of mean values of fabric self-cleaning action under five concentrations presented in table (2a) elaborates highly significant differences. The best value is obtained for C5 as 41.23 h, while the respective mean value for C4, C3, C2 and C1 are 41.61, 42.16, 42.57and 42.86 h. These results indicate that as the concentration of the coating is increased, the required time for the self-cleaning of fabric is less to done with dirt. But when the concentration decreases the time required for the self-cleaning of fabric is more. Self cleaning surface having a water contact angle greater than 150 degree and a very low roll off angle, water through these surfaces easily rolls off and completely cleans the surface in the process [11].

The individual comparison of mean values for fabric self-cleaning action under three particle size presented in Figure (2) elaborates highly significant differences. The best value is obtained for nano size S1 as 41.41 hours, while the respective mean value of S2 and S3 are 41.86 and 43.00 hours. According to these results it is proved that the particle size 10 ± 2 nm shows better effect with minimum self-cleaning action time. As the particle size increases the time required for self-cleaning of fabric also increases and take more time to roll off dirt as compared to S1. Metallic ions and metallic compounds display a certain degree of sterilizing effect. It is considered that part of the oxygen in the air or water is turned into active oxygen by a catalyst containing the metallic ion, thereby destroying the organic substance to create a sterilizing effect. Nano-materials possess enhanced catalytic abilities due to their highly stressed surface atoms which are very reactive. With the use of nano-sized particles, the number of particles per unit area is enormously increased [1].

The control value of cotton fabric samples which are not treated with nano particle finishes, have no particle size and have no effect of chemical concentration, shows no effect on the fabric.

Nano Coated Fabric Self-Cleaning Time to Roll off Ketchup Stain (Hours)

The statistical analysis of data regarding the fabric self-cleaning attribute under the effect of four different nano size self-cleaning chemicals, five different concentration of binder to chemicals, three sizes presented in table 3, reveals highly significant effect for nano-particle finishes (F), chemical concentrations (C) and nano-particle sizes (S) on nano coated cotton fabric self-cleaning time to roll off ketchup stain.

The individual comparison of mean values of treated fabrics self-cleaning action under four types of finishes presented in Figure (3), elaborates highly significant differences. The best performances is achieved for TiO_2 nano coating as 42.97 hours stain roll off time, while the respective mean value of F2, F4 and F3 are 44.09 , 45.00 and 45.02 hours. It is inferred that when stain of ketchup is applied on the fabric, the chemical finish Fl (TiO_2) take minimum time for the self-cleaning of fabric. When ketchup is applied for staining on fabric and the specimen is exposed to sun light for self-cleaning action, the initial action started after 6 hours exposure and with in 43 hours the process of self-cleaning is complete. For the self-cleaning of fabric the chemical nano finish F2 (ZnO) is the second best finish after F1. Self-cleaning fabrics containing TiO_2 nano particles have been developed based on their photo catalytic ability of oxidizing dirt and other contaminants. TiO_2 treated materials showed self-cleaning activity under daylight environment.

Microbial protective clothing has been prepared by coating military garments with silver nano particles. These coatings in addition make the fabric self-wound healing, water-repellent, and dirt-repellent. Research is also going on to prepare the chemical and biological protective clothing with self-decontamination and drug-delivery characters. The decontaminants like nano metal oxides (MgO, Al_2O_3) along with activated carbon and antibiotics in the form of nano coatings is being explored for this purpose [5].

The photo catalytic activity of TiO_2 and MgO nano particles can break harmful and toxic chemicals and biological agents. These nano particles can be pre-engineered to adhere to textile substrates by using spray coating or electrostatic methods. Finishing with nano particles can convert fabrics into sensor based materials [4]. Similarly there is substantial increase in self cleaning or stain release property of the nano TiO_2 treated fabrics.

Table 3. Analysis of variance for self-cleaning time of fabric for ketchup stain (H)

S.O.V	D.F	S.S	M.S	F.Value	Prob.
F	3	42.137	14.0455	4963.57	0.0000**
C	4	31.136	7.7840	2750.79	0.0000**
S	2	41.372	20.6861	7310.28	0.0000**
FxC	12	0.041	0.0057	1.53	N.S
FxS	6	0.027	0.0045	1.58	N.S
CxS	8	0.030	0.0038	1.33	N.S
Error	24	0.068	0.0028		
Total	59	115.378			C.V = 0.12%

* = Significant, Ns = Non-significant, ** = Highly Significant.

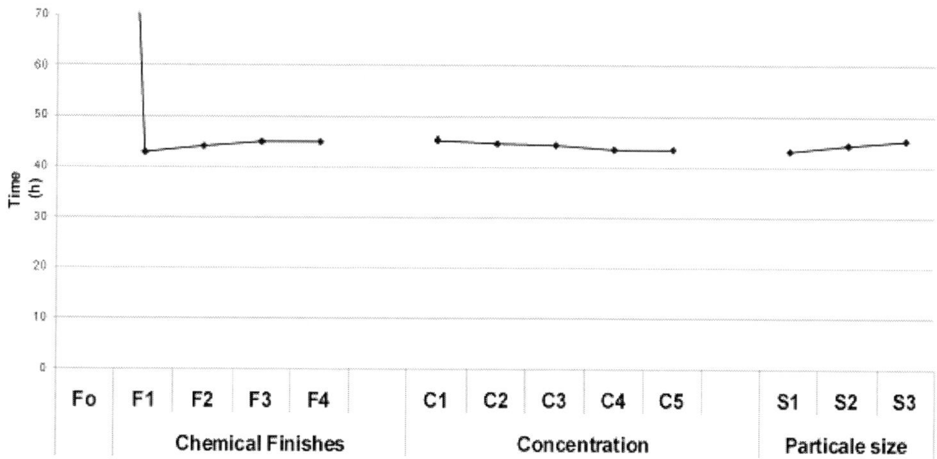

Figure 3. Graphical representation of Fabric for self-cleaning of ketchup stain (H).

It could also be seen that the self cleaning effect obtained by the nano-surface modification is independent of the nature of the substrate [material composition (cotton/blend) / fabric structure (woven/knitted)]. It has been seen that the intensity of self cleaning changes due to the changes in the size characteristics of the TiO_2 nano-particles [3].

The individual comparison of mean values of fabric self-cleaning action under five types of concentrations presented in Figure (3) elaborates highly significant differences. The best value is obtained for C5 as 43.38 hours, while the respective mean value of C4, C3, C2 and C1 are as 43.59, 44.42, 44.56 and 45.38 hours.

These results indicate that as the concentration increases, the time required for the self-cleaning of fabric is less. But when the concentration decreases the time required for the self-cleaning of fabric is more than that of C5. Self cleaning surface having a water contact angle greater than 150degree and a very low roll off angle, water through these surfaces easily rolls off and completely cleans the surface in the process. Self cleaning fabrics not only resist coffee and red wine stains but are also repellant to water, dirt, odour and are antibacterial as well [11]. One aspect of self-cleaning film that has not been well explored including film that report when they are contaminated, identify the contaminant and concentration of the contaminant, initiate a mechanism to self-clean, assess themselves as to when the film surface is restored, and stop the self-cleaning process [12]. The size and morphology of the nano particles appeared to have significant effect on the self-cleaning activity. Nano finish with 0.5 wt% nano particles and 1 wt% binder was found to provide optimum performance with respect to self cleaning activity and wash durability [13].

The individual comparison of mean values of fabric self-cleaning action under three particle sizes presented in Fig (3) elaborates highly significant differences. The best value is obtained for size 10 ± 2 nm as 43.25 hours, while the respective mean value of S2 and S3 are as 44.26 and 45.28 hours. According to these results it is conceived that minimum particle size exhibits exhibit excellent effects for self-cleaning action under minimum time. As the particle size increases the time required for the self-cleaning of fabric is increased which is not better as compared to S1. The organic pollutants decomposes to simple inorganic compound such as CO_2, H_2O and various mineral acids, which reenter the atmosphere and wash away under the effect of heat, wind or rain, so that the coating are self-cleaning with an efficiency which is dependent upon the degree of photo catalytic activity of the MO_2 catalyst, which is directly proportional to the total surface area of the MO_2 particles to which the pollutant are exposed [6]. The titanium coating on fabrics acts as an anti bacterial photo catalyst that helps to break down carbon-based molecules. Once triggered by sunlight, clothing made of such fabrics will be able to remove dirt, pollutants and micro-organism by itself [7]

The control value of cotton fabric samples which are not treated with nano particle finishes, have no particle size and have no effect of chemical concentration, shows nil effect on the fabric.

Conclusion

Different types of metal oxide nano particle finishes applied with different concentrations and nano particle sizes were applied on cotton. The effects of nano particle finishes on the different characteristics were evaluated through standard method.

It was observed that metal oxide nano particle finishes applied on cotton fabrics when exposed to sun light radiation complete its self-cleaning activity with in 48 hours.

The self-cleaning activity starts within 12 hours of exposure and roll off all the dirt and stains completely with in 44–48 hours of exposure.

During the self–cleaning process, the fabric blocks the UV light and resist against it.

This special group of Metal oxide nano particle finishes provide tendency to absorb UV light and protect the body against ultra violet rays.

When the treated fabric is spread in sunlight, the metal oxide nano particle coating absorbed maximum UV light in 48 hours, which energizes the fabric to decompose dirt and stains to help it auto clean.

The TiO_2 nano particle finishes showed comparatively batter self-cleaning effect on fabric; because it required less time for this activity where as all other finishes took more time to establish this attribute.

It was also noted that concentration with binder is directly proportional to the self-cleaning activity where as particle size is inversely associated with this activity.

Particle size and concentration of finish has more dynamic impact upon self-cleaning activity.

References

[1] Gray, R. 2006. *Self-clean technology to remove the mud, sweat and tears ofwash day for ever*: 9-15.

[2] Kathirvelu, S., L. D'Souza and B. Dhurai. 2008. Nano technology application in textiles. *Ind. J. Sci. and Tech.* 1(5):1-10.

[3] Kathirvelu, S., L. D'Souza and B. Dhurai. 2008. Multifunctional finishing of cotton and blended fabrics treated with titanium dioxide nano-particles. *Ind. J. Sci. and Tech.* 1(6): 1-9.

[4] Qian, L. and J.P. Hinestroza. 2004. Application of nanotechnology for high performance textiles. *J. Text. and App. Tech. and Mang.* 4(1):1-6.

[5] Thilagavathi, G., A.S.M. Raja and T. Kannaian. 2008. Nanotechnology and protective clothing for defence personnel. *J. Def. Sci.* 58(4): 451-459.

[6] Anonymous. 2007. Photo catalytically with wrinkle resistance. International Pub. No. WO/2006/012081-active, self-cleaning aqueous coating compositions and methods. *U.S.Patent.* 7, 261,942.

[7] Xin and Daoud. 2004. *Self-cleaning fabric plays nano magic*. [Online]. Availableat http:// www.apparelkey.com. (Verified 12 Sep. 2009).

[8] Yadav, A., V. Prasad, A.A. Kathe, S. Raj, D. Yadav, C. Sundaramoorthy and N. Vigneshwaran.2006.Functional finishing in cotton fabrics using Zinc oxide nano particles. Ind. academy of sci. *Bull. Mater. Sci.*29 (6): 641-645.

[9] Mao, Z., Q. Shi, L. Zhang and H. Cao. 2009. The formation and UV-blocking property of needle-shaped ZnO nano rod on cotton fabric. *Science Direct.*(517)8:2681- 2686.

[10] Abidi, N., E.F. Hequet and G. Abdalah. 2001. *Cotton fabric andUV-protection.* National cotton council, Memphis TN. 2(2001):1301-1303.

[11] Malik, T., S. Nogia and P. Goyal. 2009. *Self cleaning textile – an overview.* [Online]. Available at http://www.fibre2fashion.com.

[12] Skubal, R., A. L. Mcarthur and M. Newville. 2008. TiO2 Films for self-detection and decontamination. *J. Int. Photo energy*: 1-6.

[13] Gupta, K.K., M. Jassal and A. K. Agrawal. 2007. Functional finishing of cotton using titanium dioxide and zinc oxide nano particles. *RJTA*. 11(3):1-10.

[14] ASTM. 2008. Standard test method for measurement of fabric properties. ASTM D: 1424 (Vol. 7.01, 2001d), D: 2370 (Vol. 7.01, 2001b: 77-103), D: 4966, D: 6603. *Amer. Soc. for test. and mater.* Philadelphia, U.S.A.

[15] AATCC. 2001. Dimensional changes of fabrics after home laundering. Standard test method AATCC-135. *Amer. Assoc. of Text. Chem. and Color. Research Triangle Park*, North Carolina: 209-210,227-230.

[16] Faqir, M. 2004. *Statistical method and data analysis*. Kitab markaz, Faisalabad-Pakistan: 303-306.

[17] Anonymous. 2004. SPSS. *Statistical packages for social science version* 10.0. U.S.A.

In: Organic Chemistry, Biochemistry, Biotechnology ... ISBN: 978-1-62081-155-9
Editors: G. Zaikov, F. Pudel and G. Spychalski © 2013 Nova Science Publishers, Inc.

Chapter XI

Coordination of Specific Retention Volumes of Organic Compounds with Their Physical–Chemical Properties by Means of Polyparametric Equations

V. I. Rohovyk[1,*], R. G. Makitra[2], E. A. Palchykova[3], L. I. Bazylyak[2], G. G. Midyana[2] and G. E. Zaikov[4]

[1]Department of General, Bioinorganic, Physical and Colloidal Chemistry, Danylo Halyts'kyj Lviv National Medical University; Lviv, Ukraine

[2]Physical Chemistry of Combustible Minerals Department, Institute of Physical–Organic Chemistry and Coal Chemistry named after L. M. Lytvynenko, National Academy of Sciences of Ukraine; Lviv, Ukraine

[3]Institute of Geology and Geochemistry of Combustible Minerals, National Academy of Sciences of Ukraine; Lviv, Ukraine

[4]N. Emmanuel Institute of Biochemical Physics Russian Academy of Sciences, Moscow, Russia

Abstract

Specific retention volumes of different organic substances on polyester and polyethylene glycol phases can be generalized by multiparameter equations that take into account the polarity, polarizability, evaporation enthalpy, and molar volume of these substances. This approach allows one to predict the retention volumes of unstudied compounds.

* E–mail: rohovyk@ukr.net.

Keywords: Specific retention volume, multiparameter equations, correlation analysis, chromatographic phases

Introduction

In spite of the fact that chromatographic methods of analysis are of widespread occurrence, no reliable methods exist to date to correlate the chromatographic characteristics of substances, first of all, specific retention volume V_R , with their physicochemical properties.

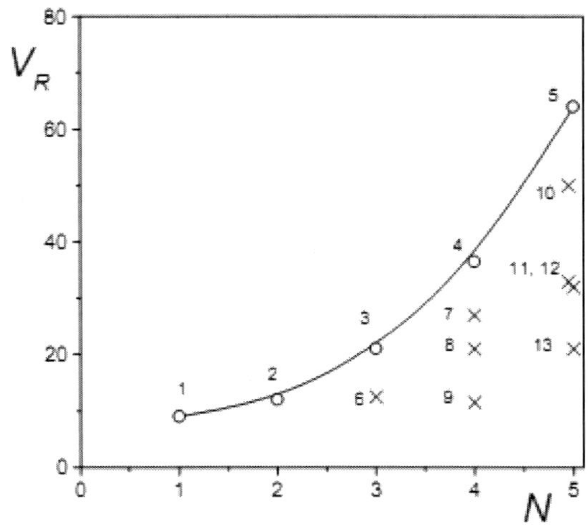

Figure 1. Dependence of the specific retention volume of alcohols V_R on the stationary phase *"diethyleneglycol adipinate/celite"* at 120 ^0C on the number of carbon's atoms N. 1 – methanol, 2 – ethanol, 3 – *n*–propanol, 4 – *n*–butanol, 5 – *n*–pentanol, 6 – *iso*propanol, 7 – *iso*butanol, 8 – butanol–2, 9 – *tert*butanol, 10 – *iso*amyl alcohol (2–methylbutanol)–1, 11 – pentanol–2, 12 – pentanol–3, 13 – 2–methylbutanol–2.

The studies known in the field [1, 2] either are qualitative in character or yield results suitable only for substances similar in structure, preferably, for homologous series. This is obviously due to the intricate character of interactions occurring in the stationary chromatographic phase–sorbate system, which cannot be adequately described by the single parameter. The attempts directly to interlink the specific retention volumes V_R and molecular weights of the substances, their molar volume V_M, evaporation enthalpy $\Delta H_{evap.}$ or *Hildebrand's* solubility parameter δ_H, were unsuccessful. *For example*, it is shown on Figure 1. the ratio between the number of carbon's atoms into molecules of lower alcohols (thereby their molecular weights) and V_R on the phase *"diethyleneglycol adipinate/celite"* at 120 ^0C.

As we can see, even in a series of the primary *n*–alkanoles an increasing of the retention volumes V_R proceeds accordingly to complicated nonlinear dependence on N; and in a case of the isomers, they are essentially deviated from this dependence sideways to decreasing. The similar dependence has been determined also for the function lgV_R. Other deviation, but sideways to increasing, is observed at the introduction of molecules by cyclic structure: *for example*, V_R for *n*–hexanol consists of 107, and for cyclohexanol – 164 under the same conditions. Similar deviations are observed also in other homologous series, *for example*, in cases of ketones, hydrocarbons and others. This is obviously that the simultaneous discussion of chemical substances by different nature is not possible.

Such type of non–agreement should be expected. As it is known, the majority of physical–chemical characteristics are not additive and the complicated dependence on the structure of substances is typical for them [3, 4]; that is why, the increments run should be taken into account at the attempts of the calculations. By example of the above–said is the work [5] accordingly to which the $\Delta H_{evap.}$ of the substances can be calculated based on their parameters σ and E_s from the *Hammet's* equation, but upon separate dependencies for ketones, alcohols or halogen derivatives.

Recently, *Abraham* and coworkers [6–8] have suggested that data on the retention of substances on stationary phases in gas–liquid chromatography *(GLC)* may be considered on the basis of the principle of linearity of free energies *(LFE)*. However, the characteristics of substances used in these works were mainly obtained by using different additive approaches.

It was of interest to study the bulk of *GLC* experimental data using the *Koppel–Palm* equation. Some characteristics of this equation are determined experimentally [9]. We showed in previous works that this equation was also effective in generalizing data on the solubility of gases and the swelling of polymers and coals in different solvents [10–12], if the factors of the medium cohesive energy and molar volume of solutions were additionally taken into account. The first factor is proportional to the squared *Hildebrand* solubility parameter δ_H^2; the molar volume of solvents V_M, reflects the effect of structural factors (equation 1):

$$lg\,Q = a_0 + a_1 f(n^2) + a_2 f(\varepsilon) + a_3 B + a_4 E_T + a_5 \delta_H^2 + a_6 V_M \quad (1)$$

In equation (1) Q is the solubility of gases in liquids (*in mole fractions*) or the mole amount of the absorbed solvent, n and ε are the refractive index and the dielectric constant of solvents under study, which determine their polarizability and polarity, respectively (that is, these factors are responsible for nonspecific solvation); and B and E_T are the basicity after *Palm* and the electrophilicity after *Reichardt*, respectively, which determine the capability of substances for acid–base interactions, that is, for specific solvation. The same equation turns out effective for generalizing data on the diffusion coefficients of organic liquids in polymers [13].

Experimental Data (from [14])

Because *GLC* is based on quasi–equilibrium processes of gas distribution between liquid stationary and gaseous mobile phases occurring under dynamic conditions, it can be expected that equations of type (1) will be suitable for generalizing data on the specific retention volumes of substances on stationary phases in *GLC* separations. To examine this hypothesis, we used the corresponding data for over sixty substances, for which V_R characteristics are published in [14]. Taking into account that the sorbates under study differ in nature and include compounds from nonpolar hydrocarbons to alcohols, which are capable of forming hydrogen bonds, this set can be considered rather representative with respect to the ability of substances to participate both in physical interactions and in the formation of hydrogen bonds or charge–transfer complexes. Retention volumes V_R are obtained on polar phases that were capable of serving as acceptors of hydrogen bonds, *namely*, esters of acids; for generalizing have been used data for polyethylene glycol too. Original data on the retention volumes V_R obtained at 120 ^0C on these stationary phases are represented in Table 1 and Table 2.

Table 1. Specific retention volumes V_R at 120 ^0C on phase «*dioctylphtalate/celite*»; experimental values $lgV_{Rexp.}$ and calculated one's $lgV_{Rcalc.}$ in accordance with the equation (4), and also their deviations $\Delta lg\ V_R$

№	Substances	V_R	lg $V_{Rexp.}$	lg $V_{Rcalc.}$	Δ lg V_R
1	hexane	11,1	1,0453	1,1920	0,1467
2	n–octane	46,4	1,6665	1,7770	0,1105
3	n–decane	177,0	2,2480	2,3051	0,0571
4	n–dodecane	645,0	2,8096	2,8005	–0,0091
5	tetradecane	2337,0	3,3687	3,2779	–0,0908
6	n–hexadecane	7800,0	3,8921	3,7567	–0,1354
7	cyclohexane	21,8	1,3385	1,3685	0,0300
8	benzene	32,5	1,5119	1,6992	0,1874
9	toluene	69,4	1,8414	1,9200	0,0786
10	o–xylene	167,0	2,2227	2,1965	–0,0262
11	m–xylene	138,0	2,1399	2,1327	–0,0071
12	p–xylene	139,0	2,1430	2,1112	–0,0318
13	ethylbenzene	124,0	2,0934	2,1031	0,0097
14	dichloromethane	14,4	1,1584	1,2017	0,0433
15	chloroform	28,2	1,4502	1,4200	–0,0303
16	tetrachloromethane	27,1	1,4330	1,5455	0,1125
17	1,2–dichloroethane	36,9	1,5670	1,5779	0,0109
18	methanol	5,1	0,7076	0,4887	–0,2189
19	ethanol	8,6	0,9345	0,9500	0,0155
20	n–propanol	18,7	1,2718	1,3224	0,0506
21	isopropanol	11,3	1,0531	1,2574	0,2043
22	n–butanol	40,9	1,6117	1,6233	0,0116
23	isobutanol	31,0	1,4914	1,5785	0,0871
24	sec–butanol	8,3	0,9191	1,5490	0,6299
25	tert–butanol	24,8	1,3945	1,4781	0,0837
26	n–amyl alcohol	85,0	1,9294	1,8936	–0,0358
27	isoamyl alcohol	31,2	1,4942	1,8814	0,3873
28	pentanol–2	49,2	1,6920	1,8328	0,1408
29	n–hexanol	40,9	1,6117	2,1560	0,5443
30	isohexanol	136,0	2,1335	2,3498	0,2163
31	n–heptanol	337,0	2,5276	2,4050	–0,1227
32	n–octanol	656,0	2,8169	2,6276	–0,1893
33	2–ethylhexanol	470,0	2,6721	2,5791	–0,0930

№	Substances	V_R	lg $V_{Rexp.}$	lg $V_{Rcalc.}$	Δ lg V_R
34	cyclohexanol	207,0	2,3160	2,2682	–0,0478
35	allyl alcohol	20,4	1,3096	1,4439	0,1342
36	2–chloroethanol	72,5	1,8603	1,6585	–0,2018
37	diethyl ether	6,7	0,8261	0,9777	0,1516
38	dipropyl ether	24,0	1,3802	1,5495	0,1693
39	isopropyl ether	12,6	1,1004	1,4691	0,3688
40	amyl ether	356,0	2,5515	2,3863	–0,1652
41	isoamyl ether	215,0	2,3324	2,3892	0,0568
42	tetrahydropyran	37,8	1,5775	1,5020	–0,0755
43	tetrahydrofuran	22,1	1,3444	1,2667	–0,0777
44	furan	8,7	0,9395	1,0156	0,0761
45	1,4–dioxane	45,0	1,6532	1,2351	–0,4182
46	dimethoxyethane	15,9	1,2014	1,2148	0,0134
47	acetone	10,4	1,0170	0,9320	–0,0850
48	methylethylketone	21,5	1,3324	1,2567	–0,0757
49	diethylketone	41,0	1,6128	1,5446	–0,0681
50	2–hexanone	81,4	1,9106	1,8130	–0,0976
51	cyclopentanone	101,0	2,0043	1,7186	–0,2857
52	cyclohexanone	210,0	2,3222	1,9976	–0,3246
53	ethyl formate	10,3	1,0128	0,9443	–0,0685
54	methyl acetate	10,6	1,0253	0,9089	–0,1165
55	ethyl acetate	18,2	1,2601	1,1972	–0,0629
56	n–propyl acetate	36,1	1,5575	1,4783	–0,792
57	n–butyl acetate	73,7	1,8675	1,7317	–0,1357
58	n–amyl acetate	148,0	2,1703	1,9934	–0,1769
59	isoamyl acetate	116,0	2,0645	1,9685	–0,0960
60	ethyl propionate	34,1	1,5328	1,4650	–0,0678
61	vinyl acetate	15,9	1,2014	1,0809	–0,1205
62	methyl acrylate	21,7	1,3365	1,2895	–0,0469
63	water	4,6	0,6628	0,4187	–0,2441
64	dibutyl ether	12,3	1,0899	2,0313	0,9414

Characteristics of the retained liquid substances were taken from [15, 16]; the calculation was performed according to the recommendations for reporting issued by the International Group for Correlation Analysis in Chemistry, *IUPAC* [17].

Results of Calculation and Discussion

The generalization of all V_R data for sixty–four substances on dioctylphtalate given in Table 1 according to equation (1) results in an expression with a relatively low value of the multiple correlation coefficient R = 0,940. The exclusion according to [17] of the most deviating data from the consideration leads, respectively, to expressions with higher values of R: after the exclusion of dibutyl ether R value growths to 0,957. Thus, it was obtained a six–

parameter equation, which correlates the logarithms of the specific retention volumes of substances with their physicochemical characteristics (according to [17] the equation is considered adequate if $R \geq 0{,}95$).

Table 2. Experimental values V_R of substances on dibutyltetrachlorophtalate (1), diethyleneglycol adipinate (2), diethyleneglycol sebacinate (3)

№	Substance	V_R		
		1	2	3
1	hexane	9,7	1,6	3,6
2	n–octane	41,8	4,9	12,5
3	n–decane	165,0	14,0	39,6
4	n–dodecane	620,0	38,4	122,0
5	tetradecane	2261,0	107,0	373,0
6	n–hexadecane	7900,0	297,0	1110,0
7	cyclohexane	19,5	4,6	8,8
8	benzene	37,6	17,0	22,5
9	toluene	80,3	28,8	42,6
10	o–xylene	212,0	62,1	98,1
11	m–xylene	163,0	48,6	79,6
12	p–xylene	168,0	47,0	77,1
13	ethylbenzene	140,0	44,0	71,5
14	dichloromethane	12,2	10,8	13,1
15	chloroform	25,2	17,6	22,9
16	tetrachloromethane	28,6	10,2	15,8
17	1,2–dichloroethane	32,7	26,6	31,7
18	methanol	4,1	9,5	8,1
19	ethanol	6,6	11,7	11,2
20	n–propanol	15,0	20,8	21,6
21	isopropanol	9,0	12,4	12,7
22	n–butanol	32,9	37,1	41,6
23	isobutanol	24,3	27,2	30,8
24	sec–butanol	11,1	20,7	23,7
25	tert–butanol	20,0	11,4	12,9
26	n–amyl alcohol	70,4	63,9	77,6
27	isoamyl alcohol	56,8	49,7	61,4
28	pentanol–2	40,7	33,9	42,4
29	n–hexanol	33,4	107,0	141,0
30	isohexanol	117,0	88,5	114,0
31	n–heptanol	287,0	182,0	253,0
32	n–octanol	560,0	321,0	477,0
33	2–ethylhexanol	400,0	231,0	343,0
34	cyclohexanol	180,0	164,0	194,0
35	allyl alcohol	16,8	29,4	27,5
36	2–chloroethanol	55,2	127,0	110,0
37	diethyl ether	6,2	53,1	3,7

№	Substance	V_R		
		1	2	3
38	dipropyl ether	21,0	481,1	—
39	isopropyl ether	11,1	—	5,3
40	dibutyl ether	85,3	2,4	31,2
41	amyl ether	332,0	5,2	106,0
42	isoamyl ether	197,0	2,8	63,8
43	tetrahydropyran	36,6	14,6	—
44	tetrahydrofuran	20,6	12,6	16,0
45	furan	9,3	6,3	7,5
46	1,4–dioxane	49,2	37,9	41,7
47	dimethoxyethane	25,0	15,6	19,3
48	acetone	9,2	9,1	9,0
49	methylethylketone	18,9	14,2	16,0
50	diethylketone	36,2	20,9	26,6
51	2–hexanone	75,1	36,5	49,3
52	cyclopentanone	92,6	77,8	84,8
53	cyclohexanone	198,0	133,0	160,0
54	ethyl formate	9,7	8,0	8,6
55	methyl acetate	17,0	—	9,2
56	ethyl acetate	33,5	11,2	19,3
57	n–propyl acetate	33,5	18,6	24,5
58	n–butyl acetate	68,8	30,9	42,8
59	n–amyl acetate	141,0	50,8	79,1
60	isoamyl acetate	110,0	39,6	61,7
61	ethyl propionate	31,1	16,3	22,3
62	vinyl acetate	15,4	11,3	13,2
63	methylacrylate	20,7	15,2	17,7
64	water	4,3	19,4	14,0

In accordance with the [17], the significance of some terms of the equation was determined by their sequential exclusion followed by the calculation of the *R* values for the obtained equation with reduced number of terms. In the case when *R* of such an equation decreases slightly, the contribution of the excluded term to the value of lgV_R is considered insignificant. In that way the insignificance of the basicity parameter *B* is determined. The effect of the δ_H^2 term is of relatively low significance too.

An analysis of obtained equation allows the conclusion that the retention time is determined by the physical interactions of absorbed substances with the stationary phase, first dipole–dipole interaction. At exclusion terms $f(n^2)$ or $f(\varepsilon)$ the value *R* becomes less than 0,95. At the same time, the insignificance of solvents electrophilicity terms shows that hydrogen bonds, for example, between dissolved molecules of alcohols and carbonyl groups of diester, are hardly formed, probably, for reasons of a relatively high temperature (120 ^0C) in the chromatographic column. Larger volumes of molecules V_M also contribute to the increase of the retention time, apparently, because of hindered diffusion to the surface of phase and evaporation to the carrier gas flow. It is surprising that the specific retention volumes of substances grow with an increase in their self–association characterized by the *Hildebrand* solubility parameter δ_H^2; however, this becomes clear if one takes into account that $\delta_H^2 = (\Delta H_{evap} - RT)/V_M$; that is it varies in parallel with substances evaporation enthalpy. As a result, the larger energy consumption for the evaporation of substances, the more difficult should be their desorption from the liquid phase film by the carrier gas flow. Therefore, we replaced the parameter δ_H^2 in equation (1) with the evaporation enthalpy $\Delta H_{evap.}$ at 298 *K*. The value $\Delta H_{evap.}$ (298 *K*) is known not for all the substances. For that reason, this parameter for some substances was determined either by recalculation ΔH_{evap} values at boiling points using the *Watson* equation or by additive approaches. After such replacement the following calculation has been done in accordance with the equation (2):

$$lgV_R = a_0 + a_1 f(n^2) + a_2 f(\varepsilon) + a_3 B + a_4 E_T + a_5 V_m + a_6 \Delta H_{evap.} \qquad (2)$$

The obtained results appear to be somewhat better than the results calculated by equation (1). For example, data on the retention of all sixty–four substances on dioctylphtalate are generalized by the six–parameter equation of type (2) with R = 0,959 (after exclusion the most deviating data for dibutyl ether from the calculation):

$$lgV_R = -3{,}14 + (11{,}88 \pm 1{,}66) f(n^2) + (11{,}57 \pm 0{,}52) f(\varepsilon) - (0{,}059 \pm 0{,}336) \times 10^{-3} B - \\ - (8{,}54 \pm 12{,}14) \times 10^{-3} E_T + (8{,}28 \pm 1{,}91) \times 10^{-3} V_M + (1{,}62 \pm 0{,}70) \times 10^{-2} \Delta H_{evap.} \qquad (3)$$

R = 0,959, $S = \pm$ 0,181, n = 63

The factors determining the retention time of substances are evaporation enthalpy and molar volume; the pair correlation coefficient r between lgV_R and V_M is equal to 0,823, and between lgV_R and $\Delta H_{evap.}$ is equal to 0,756. The effect of the other terms of the equation is relatively insignificant, because the values of r between them and lgV_R are in the range of 0,22 for B and 0,55 for $f(n^2)$. A test of the significance of individual terms of the equation by sequential exclusion followed by the determination of R for the equation with a reduced number of terms shows (as in the case of eq. (1)) the insignificance of B and E_T. Therefore, the dependence of the retention of substances on dioctylphtalate at 120 ^0C on their properties can be described with a satisfactory accuracy by the four–parameter equation (4):

$$lgV_R = -3{,}56 + (12.61 \pm 1{,}31) f(n^2) + (1{,}41 \pm 0{,}42) f(\varepsilon) + (11{,}94 \pm 3{,}030) \times 10^{-3} \Delta H_{evap.} + \\ + (9{,}42 \pm 0{,}96) \times 10^{-3} V_m \qquad (4)$$

n = 63; R = 0,958; $s = \pm$ 0,182

In Table 1, we present experimental lgV_R and the corresponding values $lgV_{Rcalc.}$, calculated by equation (4) and their difference $\Delta lgV_R = lgV_{Rcalc.} - lgV_{Rexp.}$.

The effect of polarity is of relatively low significance, because $f(\varepsilon)$ term exclusion from the equation reduces R only to 0,950. At the same time, although the retention time of substances is determined mainly by the energy consumption for the evaporation (desorption) of substances, the factor describing their ability to nonspecific interaction conventional by the polarizability has essential corrective effect on value lgV_R as $f(n^2)$ term exclusion from the equation (4) reduces R to 0,892.

To confirm the conclusions about the deciding role of V_M and $\Delta H_{evap.}$ on V_R values, we treated in a similar way in accordance with the equation (2) data for three other phases, *namely*, dibutyltetrachlorophtalate, diethyleneglycol adipinate and also diethyleneglycol sebacinate. Corresponding original data are represented in Table 2.

The values V_R of the substances on phase *"dibutyltetrachlorophtalate/celite"* for the same 64 substances are described by six–parameter equation (5) without exclusion of some data from the calculations:

$$lg V_R = -2{,}07 + (11{,}54 \pm 1{,}70) f(n^2) + (1{,}40 \pm 0{,}53) f(\varepsilon) - (0{,}034 \pm 0{,}334) \times 10^{-3} B - $$
$$- (1{,}75 \pm 1{,}24) \times 10^{-2} E_T + (6{,}77 \pm 1{,}95) \times 10^{-3} V_M + (20{,}41 \pm 7{,}15) \times 10^{-3} \Delta H_{evap.} \quad (5)$$

$R = 0{,}956;$ $s = \pm\, 0{,}186$ $n = 64$

The signs at some terms and also their validities are the same as in the previous case and the relation between V_R and physical–chemical properties of the sorbates is satisfactory described by four–parametric equation (6):

$$lg V_R = -3{,}52 + (12{,}97 \pm 1{,}36) f(n^2) + (1{,}13 \pm 0{,}44) f(\varepsilon) + (9{,}14 \pm 0{,}99) \times 10^{-3} V_M + $$
$$+ (11{,}52 \pm 3{,}40) \times 10^{-3} \Delta H_{evap.} \quad (6)$$

$R = 0{,}955;$ $s = \pm\, 0{,}188$

The effect of polarity is of relatively low significance too, the exclusion of this term from the equation reduces R to 0,950.

Two following considered ester phases, *namely*, diethyleneglycol adipinate and diethyleneglycol sebacinate (*Table* 2) unlike to esters of phthalic acids and monooxy alcohol have unbound border hydroxy groups in their structure. However, data on V_R for the same set of substances both is satisfactory described by the similar six–parameter equation and the signs of significant terms in these equations are similar too. But, in the case of diethyleneglycol adipinate in order to obtain the dependence of value R it was needed to exclude from the consideration the data for three substances, *namely*, dioxane, *tert*–butanol and water. In a case of diethyleneglycol sebacinate the data for dioxane were excluded from the consideration.

We obtain the following equations for diethyleneglycol adipinate:

$$lg V_R = -4.20 + (13{,}86 \pm 1{,}45) f(n^2) + (2{,}91 \pm 0{,}54) f(\varepsilon) + (0{,}224 \pm 0{,}317) \times 10^{-3} B - $$
$$- (6{,}557 \pm 10{,}523) \times 10^{-3} E_T + (24{,}70 \pm 5{,}21) \times 10^{-3} \Delta H_{evap.} + (1{,}888 \pm 1{,}478) \times 10^{-3} V_M \quad (7)$$

$n = 59;$ $R = 0{,}954;$ $s = \pm\, 0{,}159$

$$lg V_R = -4{,}47 + (14{,}09 \pm 1{,}29) f(n^2) + (2{,}95 \pm 0{,}42) f(\varepsilon) + (21{,}71 \pm 2{,}77) \times 10^{-3} \Delta H_{evap.} + \\ + (2{,}77 \pm 0{,}88) \times 10^{-3} V_M \quad (8)$$

$n = 59;$ $R = 0{,}953;$ $s = \pm\, 0{,}161$

For diethyleneglycol sebacinate the following equations are obtained

$$lg V_R = -4{,}55 + (13{,}86 \pm 1{,}41) f(n^2) + (1{,}85 \pm 0{,}47) f(\varepsilon) + (0{,}059 \pm 0{,}303) \times 10^{-3} B + \\ + (15{,}833 \pm 10{,}358) \times 10^{-3} E_T + (11{,}79 \pm 6{,}19) \times 10^{-3} \Delta H_{evap.} + (6{,}943 \pm 1{,}721) \times 10^{-3} V_M \quad (9)$$

$n = 61;$ $R = 0{,}958;$

the pair correlation coefficient r between lnV_R and V_M $r = 0{,}853$; $s = \pm\, 0{,}151$

$$lg V_R = -3{,}76 + (12{,}51 \pm 1{,}13) f(n^2) + (2{,}08 \pm 0{,}38) f(\varepsilon) + (20{,}07 \pm 2{,}89) \times 10^{-3} \Delta H_{evap.} + \\ + (4{,}727 \pm 0{,}847) \times 10^{-3} V_M \quad (10)$$

$n = 61;$ $R = 0{,}956;$ $s = \pm\, 0{,}154$

The special attention should be paid into fact, that the data obtained for the esters of diethylene glycol and dicarbonic acids at 140 ^0C can be also generalized in a similar way.

It is seen that the same parameters are significant in all investigated cases; not only do their signs remain the same, but the values of the coefficients are close. The enthalpy of evaporation ΔH_{evap} is the determining factor in all the cases. Fact, that for the all investigated four phases the similarity is observed can be easy explained if to take into account that the values V_R for them are about proportional (though the exclusions can take place).

Conclusion

The conducted calculations show that the multiparameter equations of types (1) or (2) allow not only the data on the retention times of substances of different nature on the polyester phase to be adequately generalized using one equation but also the specific retention value for other unstudied substances may be predicted on the basis of their physicochemical characteristics with satisfactory adequacy. Besides, obtained generalizing equations permit to do defined conclusions about the nature of processes proceeding at the absorption–desorption of substances on stationary phase. Certainly, the proposed generalization procedure and the conclusions based on the data on the retention times for phases of similar structure need an examination using experimental data on other phases different in chemical nature to estimate the possible effect of the donor–acceptor interactions, including the formation of hydrogen bonds, on the retention times.

References

[1] Sakharov V. M., *J. analyt. chem*, (*russ. version*), 27, 1002, (1972).
[2] Kejko V. V., Prokopjev B. V., Kuz'menko L. P., Kalinina N. A., Husarov A. V., Trofimov B. A., *Izv. AN USSR, Ser. khim. nauk*, 4, 755, (1969).
[3] R. C. Reid, T. K. Sherwood, *The properties of gases and liquids. Their estimation and correlation,* 2^{-nd} Ed. Mc…Comp., № 4, S. Francisko, (1966).
[4] Yavorov S. S. *Metody rashchota phisiko–khimicheskikh svojstv uglevodorodov;* Khimiya, M.: (1978).
[5] Makitra R. G., Poluzhyn I. P., *J. of general chem. (russ. version)*, 75, 783, (2005).
[6] Kollie T. O., Pool C. F., Abraham M. H., Whiting G. S., *Analyt. Chim. Acta.*, 259, 1, (1992).
[7] Abraham M. H., Whiting G. S., Doherty R. M., Shuely W. J., *J. Chem. Soc., Perkin Trans.* p. 2, 451, (1990).
[8] Abraham M. H., *Analyt. Chem.*, 69, 613, (1997).
[9] Koppel I. A., Palm V. A., *In: Advances in Linear Free Energy Relationships*. Ed. Chapman N. B., Shorter J. Plenum Press, London, New York: 203, (1972).
[10] Makitra R. G., Pyrig Ya. N., *React. sposobn. org. soed.*, 24, 38, (1987).
[11] Makitra R. G., Zagladko E. A., *J. of physical chemistry (russ. version)*, 76, 1797, (2002).
[12] Makitra R. G., Prystansky R. E., *Khimiya tviordogo topliva*, 5, 3, (2001).
[13] Makitra R. G., Zagladko E. A., Midyana G. G., Protsajlo L. V., *J. of physical chemistry (russ. version)*, 75, 2283, (2001).
[14] Korol' A. N. *Niepodvizhnyje phazy v gazozhydkostnoj khromatografiji,* Khimiya, M.: 149, (1985).
[15] Makitra R. G., Pyrig Ya. N., Kivelyuk R. B., *Avail. VINITI Deposited Doc.*, № 628, B 86, (1986).
[16] Abboud J. L. M., Notario R., *Pure Appl. Chem.*, 71, 645, (1999).
[17] Recommendation for Reporting of Correlation Analysis in Chemistry using Regression Analysis, *Quant. Struct. Act. Relat*, 4, 29, (1985).

In: Organic Chemistry, Biochemistry, Biotechnology ... ISBN: 978-1-62081-155-9
Editors: G. Zaikov, F. Pudel and G. Spychalski © 2013 Nova Science Publishers, Inc.

Chapter XII

Microbiological Corrosion of Metals and Protection against It

Elena L. Pekhtasheva*, Anatoly N. Neverov, Irina G. Shishkina, and Gennady E. Zaikov[1]

G.V. Plekhanov Russian Economic University, Moscow, Russia
[1]N.M Emanuel Institute of Biochemical Physics, Russian Academy of Sciences, Moscow, Russia

Abstract

Microorganisms can impact on metals differently. First of all, corrosion can be involved by aggressive metabolites of microorganisms – acids, bases, enzymes, etc. They form the corrosion-active medium, in which corrosion proceeds according to common laws of electrochemistry. Colonies of microorganisms may form growths of mycelium or mucus on the metal surface, under which due to electric potential difference in different areas of metal surface and metal ion assimilation by microorganisms pitting corrosion develop.

Keywords: Corrosion, metals, microorganisms, protection, aggressive metabolites

Biodamages of metals are commonly called microbiological corrosion (biocorrosion) of metals. In the everyday life, the cases of biocorrosion of metals are less frequent, than the cases of nonmetal material biodamages. Metals themselves are higher biostable materials, and some of them are biocides. In machines, devices and other technical products, metals are usually used with various protective and decorative paint and other coatings, which are first subject to the impact of biodamaging agents and preserve metal from biocorrosion. The external manifestations of biocorrosion are similar to common corrosion accompanied by rust formation [1, 2].

* E-mail: pekhtashevael@mail.ru.

Microorganisms can impact on metals differently. First of all, corrosion can be involved by aggressive metabolites of microorganisms – acids, bases, enzymes, etc. They form the corrosion-active medium, in which corrosion proceeds according to common laws of electrochemistry.

Colonies of microorganisms may form growths of mycelium or mucus on the metal surface, under which due to electric potential difference in different areas of metal surface and metal ion assimilation by microorganisms pitting corrosion develop.

Curious Facts

American Gold Mining Corporation in Denver (Colorado) uses sulfobacterium of *Thiobacillus* genus for metal mining from sulfide ore. Resulting the vital activity of these microorganisms, gold concentration in the ore increases. This biomethod of mining does not allow sulfur oxides release to the atmosphere that eliminates acid rains and environmental pollution. French scientists expect to use this bacterium for ecologically friendly recovery of biocobalt from copper mining wastes.

Biodamaging of metals induced by microorganisms may proceed in different ways:

- By means of direct impact of microorganism metabolites on metal;
- Via organic products formation, which may act as depolarizers or catalysts of corrosive reactions;
- By way, when corrosive reaction are a separate part of the metabolic cycle of bacteria.

As concerns bacteria, corrosion of metals is most often associated with the activity of sulfate-reducing bacteria; thionic bacteria, which oxidize sulfur and sulfur compounds to sulfuric acid; iron bacteria, which oxidize ferrous iron to ferric iron.

Sulfate-reducing anaerobic bacteria are biocorrosion initiators. They are able to convert metal sulfate compounds to sulfides formed in the reaction of hydrogen sulfide released in this process with metals.

There are several hypotheses about the mechanism of anaerobic corrosion of steel, iron, aluminum and their alloys under the effect of sulfate-reducing bacteria.

One of the hypotheses suggests that at high iron sulfide concentration in the medium, it forms a galvanic couple with iron, where sulfide plays the role of cathode, and iron being the anode endures corrosion.

Thionic bacteria oxidize sulfides and other reduced sulfur compounds to sulfates. The rate of bacterial oxidation of sulfides may be millions of times higher than the rate of common chemical oxidation. As a result, sulfur acid is formed in large amounts and very quickly, and this forms the aggressive corrosive medium.

Curious Facts

When constructing the underground in Kiev (Ukraine), in some sections neutral ground water flowing about the tube turned to 0.1 n sulfuric acid due to thionic bacteria multiplication within several months and corroded the steelwork of the tube by 40%.

Iron bacteria corrode metal surfaces contacting with water. They form greasy accumulations not washed off by water on weld joints and other metal surfaces. Non-water-washed areas are formed beneath them. Therefore, these areas are poorly aerated, have lower potential and act as anodes. In the anodic zone iron dissolves and corrosion happens.

Some scientists suggest that gold deposits might be formed due to microorganisms' action. In laboratory, some particular cenoses of microorganisms and algae inhabiting in the Sea of Okhotsk were cultivated during two month on culture media containing dissolved and suspended (colloid) gold. As shown by spectral analysis, these cenoses transfer gold from solution and suspensions into sediments. Gold particles are 3-9 μm in size and form 35-70% of the sediment. Thus, it is shown that microorganism cenoses may participate in enlarging gold particles and concentrating them in deposits.

Curious Facts

In June, 1999, in Saint-Petersburg (Russia), an entrance screen of Sennaya Ploschad' underground station was broken. On its structures, metabolites of thionic and nitrifying bacteria were detected.

A group of scientists has proven the presence of widely spread processes of construction materials and entire engineering and technical facilities biodamaging in Saint-Petersburg. Under conditions of increased humidity microbes intensively colonize the human environment – buildings, utilities, various goods and other material objects.

The problem of metal damaging by fungi is the most poorly studies, because until recently, it has been suggested that biodamages of metals are mostly induced by bacteria [1-14].

However, fungal corrosion of metals does exist and in a number of cases, it makes not lesser harm to metal structures, than bacterial corrosion. Fungi retain moisture on the metal surface and excrete organic acids and, thus, provide corrosion of parts from brass, steel, aluminum and its alloys. Microbiological corrosion products, as well as fungus mycelium forming bridges between metal contacts of articles promote occurrence of electrolytes on the surface of contacts and cause short circuiting of electric networks or deterioration of electric parameters of the articles.

Massive losses due to corrosion after tests performed during 12 day in the presence of A. niger reached 4 g/m^2 for aluminum, 18 g/m^2 for copper and 33 g/m^2 for iron, which is 4 times greater than losses of each metals caused by common corrosion. The change of physicochemical properties of the medium in the course of metabolism proved by changes in pH, redox potential of the medium and electrochemical potentials of metals is the basic factor that induces corrosion of metals in the presence of fungi.

The tests of electrolytic copper plates agglutinated and annealed under different conditions in the salt medium, after applying A. flavus and A. niger fungus conidia suspension on their surface, have indicated growth of fungi and plate damaging in all alternatives of the tests. In the studied under tropical climate conditions, damaging of some kinds of powder materials and bimetal wire by fungi was observed.

It is suggested that the primary mechanism of damaging is fungus hyphae intrusion into particular areas of the sample surface, and the secondary mechanism is the action of their metabolites on the metal surface.

Among tested samples of metals and alloys, the highest funginertness is demonstrated by high strength carbon steel and aluminum-magnesium alloy. Pure aluminum endures the highest corrosive changes.

Biostability of metals is evaluated by visual inspection of corrosion, the area of corrosive damages, mass loss of the samples (after corrosion product removal), and corrosive damage depth.

Scanning electron microscopy tests indicated that living cells of fungi, yeasts and bacteria, spores, dead cells of microorganisms, cellular debris and various inorganic substances are included in the structure of biodamages of metals.

Biological fouling is tightly fixed on the metal surface. In places of fouling localization, steel, aluminum and copper have various biodamages: from microcracks and microcraters to complete destruction of metal in these areas.

On the surface of pure uncontaminated metal having no contact with organic materials, e.g. lubricants, polymeric films or paints, fungi may not develop. In this connection, biocorrosion of metals induced by fungi is some sort of secondary damage. They first colonize and propagate on organic materials contacting with metal, and then mycelium extends to metal and initiates corrosion caused by its metabolites – acids and enzymes.

Biocorrosion was observed on threaded joints, electric terminals, etc. Such damages are typical of electronic and optical devices. Mycelium formation on the surface of electric terminals of the devices induced failures of the entire device due to short circuiting or electric circuit break caused by corrosion product nodes on the surfaces.

Protection methods for metals against biocorrosion are based on the application of chemical biocides, as well as on reasonable selection and the use of biostable materials in technical appliances. Compliance with hygienic rules in production and operation of mechanisms is of great significance.

Protection of metals against biocorrosion is performed by the same biocides applied to protection of nonmetal materials. A significant requirement to these biocides is that they must not show aggression to metals and cause their corrosion, because some biocides are dangerous in this aspect.

References

[1] E.I. Andreyuk, V. Bilan, E.Z. Koval, Microbial corrosion and its agents, *Naukova Dumka*, (Publishing House of Ukrain Academy of Sciences), Kiev, 1980, 258 pp. (in Rus.).

[2] R. Blagnik, B. Newly, Microbiological corrosion, *Khimiya (Chemistry)* Publishing House, Moscow, 1965, 222 pp. (in Rus.).
[3] E.L. Pekhtasheva, Topical issues of biological damage, *Nauka (Science)* Publishing House, Moscow, 1983, 265 pp. (in Rus.).
[4] A.N. Neverov, Actual problems of biological damage and protection of materials, components and structures, *Science*, Scientific Council for biological damage of the USSR Publishing House, Moscow, 1989, 256 pp. (in Rus.).
[5] A.N. Neverov, *Biodamage and methods for assessing the biological stability of materials,* Scientific Council for biological damage of the USSR Publishing House, Moscow, 1988, 140 pp. (in Rus.).
[6] A.N. Neverov, *Biodamage, methods of protection*, The Scientific Council for biological damage of the USSR Publishing House, Poltava, 1985, 182 pp. (in Rus.).
[7] I.G. Kanev, Biological damage of industrial materials, *Nauka (Science)* Publishing House, Leningrad, 1984, 232 pp. (in Rus.).
[8] *Biological problems of environmental materials*, Ed. by A.N. Neverov, Scientific Council on the biological damage of Sciences Publishing House, Penza, 1995, 108 pp. (in Rus.).
[9] *Methods for determining the biological stability of materials*, Ed. by N.A. Plate, Science Council of biological damage of the USSR Publishing House, Moscow, 1979, 202 pp. (in Rus.).
[10] Microorganisms and lower plants - the destroyers of materials and products, Ed. by N.A. Plate, *Nauka (Science)* Publishing House, Moscow, 1979, 225 pp. (in Rus.).
[11] First Conference on Biological Damage, Ed. by N.A. Plate, *Nauka (Science)* Publishing House, Moscow, 1978, 226 pp. (in Rus.).
[12] Problems of biological damage to materials. *Environmental aspects* Ed. by A.N. Neverov, Scientific Council of biological damage of the USSR Publishing House, Moscow, 1988, 124 pp. (in Rus.).
[13] Environmental problems of biodegradation of industrial, *construction materials and industrial wastes,* Ed. by A.N. Neverov, Scientific Council on Problems of biological damage Publishing House, Penza, 2000, 192 pp. (in Rus.).
[14] *Fourth Conference on Biological Damage*, Ed. by N.A. Plate, Scientific Council of biological damage of the USSR Publishing House, Nizhnii Novgorod-city, 1991, 99 pp. (in Rus.).

In: Organic Chemistry, Biochemistry, Biotechnology ... ISBN: 978-1-62081-155-9
Editors: G. Zaikov, F. Pudel and G. Spychalski © 2013 Nova Science Publishers, Inc.

Chapter XIII

Biodamaging and Protection of Paint Materials

Elena L. Pekhtasheva*, Anatoly N. Neverov, Irina G. Shishkina and Gennady E. Zaikov[1]
G.V. Plekhanov Russian Economic University, Moscow, Russia
[1]N.M Emanuel Institute of Biochemical Physics,
Russian Academy of Sciences, Moscow, Russia

Abstract

The paper is including information about biodegradation, biodamages and biodeterioration of paint materials by *Aspergillus, Penicillium, Fusarium, Trichoderma, Alternaria, Cephalosporium, Pullularia* genus fungi and *Pseudomonas, Flavobacterium genus bacteria. The contributors are discussing the ways for stabilization of paint materials.*

Keywords: Biodegradation, biodamages, biodeterioration, protection, paint materials, stabilization, bacteria

Paint coatings are applied for protective and decorative purposes. They fashion goods nicely, make their keeping simpler, protect against degradation, increase fireproof, chemical resistance, etc.

Paint materials are divided into basic (drying oils, lacquers, paint compositions) and auxiliary (primers, paste fillers, etc.).

Paint materials are based on film-forming substances (film-forming agents). Moreover, they may include pigments, dyes, thinners and diluters, plasticizers, drying catalysts, curing agents and other additives.

* E-mail: pekhtashevael@mail.ru.

The film-forming substances (oils, resins, cellulose ethers) are able to form a solid glossy film on the surface and bind and fix particles of the composition components, e.g. a pigment in paints. Film formation may proceed due to thinner or dispersion medium evaporation (for alcoholic varnishes, nitrolacquers, water-based paints) or as a result of chemical transformations, accompanied by chain molecule cross-linking and polymer formation (for polyester and epoxy varnishes). Liquid film-former conversion to a solid film is induced by curing agents, heat, oxygen and other factors.

Paint materials and coatings applied under conditions favorable for growth and development of molds, bacteria and other microorganisms and being the nutrient substrate for these agents of biodamages, may be subject to microbiological damages. When applied under natural conditions, biodamaging of paint coatings resulting overgrowth by living organisms and aggressive action of their metabolites (enzymes, organic acids) is usually combined with the impact of external physical and chemical factors (solar radiation, increased humidity, temperature, etc.), which induce aging of materials. The latter promotes development of biodamages. Aging and biodamaging processes may proceed simultaneously or be mistimed, but in most cases, they complement each other, accelerating and aggravating deterioration of materials and worsening their operation and decorative properties. Under real conditions, it is difficult to determine, to what extent the damage of paint coating is due to biological or physicochemical factors [1-51].

Damages of paint materials (especially water-based paints) cause by microorganisms that happens during manufacture (storage) and damages of cured protective and decorative coatings applied to support should be distinguished. In the first case, microorganisms usually get into semi-products of lacquers and paints during technological cycle of their manufacture (with raw materials, water, from the air of manufacturing area, etc.); in the second case, spores of fungi and bacteria precipitate on ready coating surfaces from the surrounding, which is to a great extent promoted by various contaminations.

It should be outlined that biostability of paint coatings highly depends on the origin and properties of protected materials (the support) and, as found in a number of cases, increases in the following sequence: wood, metal, bricks and other construction materials. Meanwhile, coatings applied on nonferrous metals are destroyed somewhat faster that those applied on ferrous metals.

Sometimes, even insignificant quantities of organic matter present in air, and contamination of equipment and surfaces of manufacturing areas may be the source of carbon necessary for microorganisms' nutrition. This increases the probability of painted surface biodamaging on equipment located in the premises and premises themselves. This takes place, e.g., in the enterprises of microbiological, baking, beef and dairy, and chemical industry, in some warehouses, etc. This necessitates increasing of hygienic requirements and the use of paint materials with increased fungal resistance.

Cracking and exfoliation of protective paint coatings, crocking and piling, formation of spots and through punctures are typical signs and manifestations of the destructive activity of biological agents intensified by physical, chemical, mechanical and other adverse external impacts.

Sometimes, surface staining of paint coatings (grey, black, violet, pink, etc.) caused by microorganisms' pigments may be erased or washed off. But if the damage of coating materials is bulky (mycelium hyphae germination), damages cannot be removed in this manner, and decorative and protective properties of the coating may not be restored. Cracks, exfoliation and swelling of paint coatings may be induced by microorganisms present on the surface of beneath the coating, on the protected surface. Fungus growth and development beneath the coating is accompanied by gas evolution and swelling of the coating.

Mold fungi are the main agents of microbiological damages of paint coatings. Bacterial damages are less frequent observed. They are characterized by colorless or stained mucous bloom formation. Beneath the paint coating, microbiocenoses of a complex composition, which include bacteria and fungi, are met. Among microorganisms damaging paint coatings, *Aspergillus, Penicillium, Fusarium, Trichoderma, Alternaria, Cephalosporium, Pullularia* genus fungi and *Pseudomonas, Flavobacterium genus bacteria* are detected. Coating damage by fungi is caused either by components in the coating structure, or by surface contaminants under the effect of metabolites excreted by mycelium propagating due to coating contaminants.

Typical signs of biodamage of paints themselves may be a change of their color, gas evolution (package swelling), occurrence of foreign odor, fluidifying and, finally, gelation.

Biostability of finished paint coatings depends, to a considerable extent, on their composition, chemical nature of polymeric binder and pigments. Biostablity is also affected by other components of varnishes and paints (thinners, diluters, stabilizers, curing agents, etc.).

Film-forming substances generally define biostability of paint materials and protective coatings on their basis. Here the decisive factor, on the one hand, is chemical structure of polymeric film-former and, on the other hand, its physical properties in both uncured and cured state (swelling capacity, moisture capacity, hardness, surface smoothness, porosity, etc.).

The increased drying rate of the film-forming substance is favorable for providing biostability of the protective coating. The lower the moisture amount absorbed during curing is, the lesser probability of mold development in future is. Fungal resistance increase is promoted by the use of film-forming substances making smooth flat glossy films, which surface is worse contaminated because of the absence of inequalities and roughness.

Among natural film-formers, the most widespread are vegetable drying oils (linen, cotton, hemp, sunflower, etc.). All of them are comparatively low fungus resistant. Their biostability is increased by refining, i.e. removal of water, protein products and other admixtures. The typical signs of vegetable oil damaging by microorganisms are: viscosity reduction, acidity increase, polymerization ability decrease, i.e. the film formation rate. One of the most biostable natural film-forming substances for paint materials is wood rosin. Its biostability is associated with the presence of terpenes in its composition, which possess fungicidal properties, and acid products formation in the film during protective coating formation.

Synthetic, film-forming polymers (heat-sensitive and thermoreactive ones) are less prone to microorganism damages than natural ones. Funginertness of these coatings decreases in the following sequence: epoxy, polyurethane, melamine alkyd, organosilicon, pentaphthalic.

Bitumen used as film-forming agents have insufficient biostability. To increase biostability, phenolic, maleic and other synthetic resins are added to bitumen paints and protective coatings.

Thermoplastic synthetic resins based on chlorinated rubber, styrene-butadiene and vinyl chloride-vinyl acetate copolymers used for production of quick-drying lacquers and non-fouling paint coatings are characterized by high biostability.

Polyvinyl acetate dispersion is the widespread polymeric binder. Paints, coatings, mastics, primers and other materials on its basis are non-funginert. Non-biostable are not only polyvinyl acetate paint coatings, but also liquid paints themselves, which are damaged by fungi and bacteria during storage; meanwhile, their viscosity is reduced, gaseous products are formed, etc.

High resistance to microorganism degradation of thermoreactive resins applied as hot curing film-forming varnishes and enamels is explained formation of low-permeable solid smooth films. Some thermoplasts manifest fungicidal properties (phenolic plastics, amino plastics, glyphtal resins, etc.). Among resins of this group, as biostable protective covers phenolic, glyphtal, epoxy, urea-formaldehyde and melamine-formaldehyde, silicone resins and their mixture are used.

Modification of polymeric thermoreactive resins (glyphtal, phenolic, etc.) by drying oils and fatty acids improves technological properties of obtained fatty varnishes and hot-drying enamels and reduces funginertness of protective coatings because of comparatively low mold funginertness of modifying components.

Water-soluble film-forming substances, such as water-soluble cellulose derivatives, protein compounds, etc. (gums, dextrin, gelatin, albumin, casein, etc.), may be damaged by microorganisms consuming carbon from these polymers and a nutrition. Cases of microbiological damage of casein, glutinous emulsion and other paints are known. Reduced biostability of water-soluble film-forming organic agents also results from high hygroscopic property and ability of film coatings on their basis to swell in water.

Inorganic binders applied in silicate paint production, e.g. soluble glass, are characterized as funginert.

Pigments are the most important components, on which biostability of the paint coating depends. Pigments impart required color and tinting power to the paint, regulate viscosity, improve solar radiation resistance and water stability of the coating. Due to increased hardness, pigment particles mechanically hinder mycelium growth and development. They may also be toxic for mold fungi and other microorganisms.

Zinc oxide, copper oxide (I), barium metaborate and some other pigments are fungicides and, therefore, increase biostability of paint coatings.

Comparative studies of the funginertness of a number of pigments indicate that paint coatings containing zinc oxide and titanium oxide are most funginert.

At the same time, such pigments as chalk, chrome yellow, titanium dioxide, aluminum powder, chromium oxide, carbon black do not possess biocidal properties themselves; however, oily paints on their basis are highly funginert. Lower funginertness is demonstrated by oily paints with antimony and lead oxide pigments. A number of inorganic pigments and fillers, e.g. talcum, graphite and muscovite mica, reduce funginertness.

Paint coatings must be protected against microorganism damage already at the production stage. This particularly concerns water-soluble paints. Keeping production areas and warehouses clean is the necessary condition for prevention of microbiological contamination of paints and their raw materials.

Raw materials, half-finished products and auxiliary materials may occur contaminated by microorganisms, what is shown up by their staining, occurrence of foreign putrefactive odor and gas evolution. Such materials must not be used in production of varnishes and paints.

Beside prophylactic protective measures against biodamages, paint materials and coatings are protected by biocides, which are divided into two groups by application staging: biocides protecting raw materials and materials used in paint production and biocides protecting paint coatings themselves. Biocides protecting paint materials against damages during their production shall meet a number of technical requirements: preserve biocidal activity in a wide pH range for a long time, be low toxic for people, be universal for various paint materials, not change protected product color and its physicochemical properties, not deteriorate technological properties of paint materials, coating quality and adhesion to protected surface, etc.

In case of increased danger of microbiological damages, it is recommended to use preservative treated paints. Protective decorative and insulating paint coatings with added biocides are recommended for use in some kinds of electronic facilities, opto-mechanical and other devices, especially delivered to the countries with tropical climate. Preservative treated paints are applied to painting compartments with increased humidity and temperature (pools, saunas, some food industry enterprises, etc.), and to protect wood structures.

The following substances may be used as biocides for general purpose paint coatings: 1) inorganic pigments – zinc oxide, copper oxide (I), barium metaborate, etc.; 2) organic fungicides – copper 8-oxyquinolate (adheres yellowish-green to brown stain to paint coatings, may be used in the food industry due to low toxicity), salicyl anilide, Bromtan, *n*-nitrophenol, tetra- and pentachlorophenol, phthalan (trichloromethyl thiophthalimide), etc.; 3) organometal fungicides – organotin (hexabutyldistannoxane, tributyltinacrylate), organoarsenic (chlorophenoxasine), organomercury (phenylmercuric oleate, etc.), the latter are highly volatile and toxic for man, so their application is restricted [52-60].

Self-disinfecting paints are also used for finishing premises in health care and children's institutions, in food industry enterprises, on transport and in other public areas, i.e. in places, where potential hazard of occurrence and dissemination of infectious diseases does exist. Bactericidal and fungicidal properties of such paints preserved for more than 2 years is provided by injection of 2,3,5,6-tetrachloro-4-(methylsulfonyl) pyridine as a biocide in them.

References

[1] M.N. Rotmistrov, P.I. Gvozdyak, S.S. Stavskaya, *Microbial degradation of synthetic organic compounds*, Naukova Dumka Publishing House, Kiev, 1975, 224 pp. (in Rus.).

[2] R. Blagnik, B. Newly, *Microbiological corrosion*, Khimiya (Chemistry) Publishing House, Leningrad, 1965, 222 pp. (in Rus.).

[3] I.A. Yermilov, *Theoretical and practical bases of microbial degradation of chemical fibers,* Nauka (Science) Publishing House, Moscow, 1991, 248 pp. (in Rus.).

[4] L.A. Wolfe, A.I. Meos, Fibers for special applications, *Khimiya Chemistry)* Publishing House, Moscow, 1971, 224 pp. (in Rus.).

[5] V.S. Bugorkov, T.A. Ageev, V.M. Galperin, The basic directions for creating photo-polymer materials and biodestruktiruemyh, *Plastics Journal,* 1991, N 9, p. 48-51 (in Rus.).

[6] O.A. Legonkova, L.A. Sukharev, *Polymer biological resistance to biodegradation*, RadioSoft Publishing House, Moscow, 2004, 272 pp. (in Rus.).

[7] K.E. Perepelkin, *The present and future man-made fibers*, LegPromBuzines Publishing House, Moscow, 2000, № 8 (22), № 9 (23), № 10 (24), № 11 (25), № 12 (26), (in Rus.).
[8] N.M. Emanuel, A.L. Buchachenko "*Chemical physics of degradation and stabilization of polymers*", VSP International Science Publ., Utrecht, 1982, 354 pp.
[9] A.F. Bochkov, G.E. Zaikov "*Chemistry of the glycosidic bonds. Formation and cleavage* ", Pergamon Press, Oxford, 1979, 210 pp.
[10] S.D. Razumovskii, G.E. Zaikov "*Ozone and its reactions with organic compounds*", Elsevier, Amsterdam, 1984, 404 pp.
[11] N.M. Emanuel, G.E. Zaikov, Z.K. Maizus "*Oxidation of organic compounds. Medium effects in radical reactions* ", Pergamon Press, Oxford,1984, 628 pp.
[12] V.A. Afanasiev, G.E. Zaikov "*In the realm of catalysis*", Mir Publishers, Moscow, 1979, 220 pp.
[13] Yu. V. Moiseev, G.E. Zaikov "*Chemical resistance of polymers in reactive media*", Plenum Press, New York, 1987, 586 pp.
[14] G.E. Zaikov, A.L. Iordanskii, V.S. Markin "*Diffusion of electrolytes in polymers*", VNU Science Press, Utrecht, 1988, 328 pp.
[15] K.S. Minsker, S.V. Kolesov, G.E. Zaikov "*Degradation and stabilization of polymers on the base of vinylchloride*", Pergamon Press, Oxford, 1988, 526 pp.
[16] R.M. Aseeva, G.E. Zaikov "*Combustion of polymer materials*", Karl Hanser Verlag, Munchen, 1986, 389 pp.
[17] A.A. Popov, N.A. Rapoport, G.E. Zaikov "*Oxidation of stressed polymers*", Gordon and Breach, New York, 1991, 336 pp.
[18] A.F. Bochkov, G.E. Zaikov, V.A. Afanasiev "*Carbohydrates*", VSP Science Press VB, Zeist-Utrecht, 1991, 154 pp.
[19] V.A. Afanasiev, G.E. Zaikov "*Physical methods in chemistry*", Nova Science Publ., New York, 1992, 180 pp.
[20] I.N. Todorov, G.E. Zaikov, I A. Degterev "*Bioactive compounds: biotransformation and biological action*", Nova Science Publ., New York, 1993, 292 pp.
[21] V.L. Roubajlo, S.A. Maslov, G.E. Zaikov "*Liquid phase oxidation of unsaturated compounds*", Nova Science Publ., New York, 1993, 294 pp.
[22] A.L. Iordanskii, T.E. Rudakova, G.E. Zaikov "*Interaction of polymers with bioactive and corrosive media*", VSP International Publ., Utrecht, 1994, 298 pp.
[23] "*Degradation and stabilization of polymers. Theory and practice* ", Ed. by G.E.Zaikov, Nova Science Publ., New York, 1995, 238 pp.
[24] A.Ya. Polishchuk, G.E. Zaikov "*Multicomponent transport in polymer systems*", Gordon and Breach, New York, 1996, 231 pp.
[25] K.S. Minsker, A.A. Berlin "*Fast reaction processes*", Gordon and Breach, New York, 1996, 364 pp.
[26] E.Ya. Davydov, A.P. Vorotnikov, G.B. Pariyskii, G.E. Zaikov "*Kinetic pecularities of solid phase reactions*", John Willey and Sons, Chichester (UK), 1998, 150 pp.
[27] J.N. Aneli, L.M. Khananashvili, G.E. Zaikov "*Structuring and conductivity of polymer composites*", Nova Science Publ., New York, 1998, 326 pp.
[28] K.Z. Gumargalieva, G.E. Zaikov "*Biodegradation and biodeterioration of polymers.Kinetical aspects* ", Nova Science Publ., New York, 1998, 210 pp.

[29] S.K. Rakovsky, G.E. Zaikov "*Kinetics and mechanism of ozone reactions with organic and polymeric compounds in liquid phase*", Nova Science Publ., New York, 1998, 345 p.
[30] S.M. Lomakin, G.E. Zaikov "*Ecological aspects of polymer flame retardancy*", VSP International Publ., Utrecht, 1999, 158 pp.
[31] K.S. Minsker, G.E. Zaikov "*Chemistry of chlorine-containing polymers: synthesis, degradation, stabilization*", Nova Science Publ., New York, 2000, 198 pp.
[32] A. Jimenez, G.E. Zaikov "*Polymer analysis and degradation*", Nova Science Publ., New York, 2000, 287 pp.
[33] G.V. Kozlov, G.E. Zaikov "*Fractal analysis of polymers*", Nova Science Publ., New York, 2001, 244 pp.
[34] G.E. Zaikov, A.L. Buchachenko, V.B. Ivanov "*Aging of polymers, polymer blends and polymer composites*", Nova Science Publ., New York, 2002, Vol. 1, 258 pp.
[35] G.E. Zaikov, A.L. Buchachenko, V.B. Ivanov "*Aging of polymers, polymer blends and polymer composites*", Nova Science Publ., New York, 2002, Vol. 2, 253 pp.
[36] G.E. Zaikov, A.L. Buchachenko, V.B. Ivanov "*Polymer aging at the cutting adge*", Nova Science Publ., New York, 2002, 176 pp.
[37] S.A. Semenov, K.Z. Gumargalieva, G.E. Zaikov "*Biodegradation and durability of materials under the effect of microorganisms*", VSP International Science Publ., Utrecht, 2003, 199 pp.
[38] G.E. Zaikov, "*Burning, destruction and stabilization of polymers*", Scientific Basis and Technology Publishing House, St. Petersburg, 2008, 422 p. (in Rus.).
[39] A.K. Mikitaev, G.V. Kozlov, G.E. Zaikov, "Polymer nanocomposites. A variety of structural forms and applications, *Nauka (Science)* Publishing House, Moscoe, 2009, 279 p. (in Rus.).
[40] A.K. Mikitaev, G.V. Kozlov, G.E. Zaikov, "*Fractal analysis of the gas transport in polymers,* Nauka (Science) Publishing House, Moscow,, 2009, 199 p. (in Rus.).
[41] S.K. Rakovsky, G.E. Zaikov "*Interaction of ozone with chemical compounds. New frontiers* ", Rapra Technology, London, 2009, 486 pp.
[42] R.P. Naumova, *Conversion study of caprolactam by bacteria*, Ed. By N.A. Plate, Kazan State Technological University Publishing House, Kazan, 1964. - P.67-69 (in Rus.).
[43] R.P. Naumova, N.G. Zakharova, S.J. Zakharova, Degradation of synthetic lactam and ε-amino acids by microorganisms, cleaning industrial waste water. *Microbiological methods of dealing with environmental pollution*, Ed. By N.A. Plate, Nauka (Science) Publishing House (Pushchino Branch), Pushchino-town, 1979. - P.70-72 (in Rus.).
[44] K. Kato, T. Fukumura, Bacterial breakdown of ε-caprolactam, Chem.and Industr. *Journal*, 1962. - N 23. - P.1146.
[45] S. Kinoshita, E. Kobayashi, H. Okada, Degradation of ε-caprolactam by Achromobacter guttatus KF 71, *Journal of Fermentation Technology,* 1973, Vol. 51, P. 719-725.
[46] T. Uemura, Autolytic enzyme assocated with cell walls of Bacillus subtilis, *Journal of Biol. Chem.,* 1966, Vol. 241. - N 15. - P.3462 - 3467.
[47] A. Roy, Destruction of caprolactam and hexamethylenediamine bacterial groups Bac.*subtilis-mesentericus, Thesis*, Institute of Colloid and Water Chemistry, USSR Academy of Sciences, Kiev, 1975. – 162 pp. (in Rus.).
[48] A. Roy, The destruction of caprolactam microorganisms in continuous cultivation. *Scientific fundamentals of water treatment*, Ed. By N.A. Plate, Naukova Dumka

(Publishing House of Ukrain Academy of Sciences), Kiev, 1976. - Issue 2. - P.152-156 (in Rus.).

[49] I.I. Shevtsova, *The destruction of microorganisms caprolactam*, Bulletin of Kiev State University, Division of Biology, Kiev, 1969. - № 11. - P.149-152 (in Rus.).

[50] T. Fucumura, Bacterial Breakdown of ε-caprolactam and its cyclic oligomers, Plant and cell. *Physiology Journal*, 1966. - Vol.7. - N1. - P. 93-104.

[51] R.P. Naumova, I.S. Belov, The transformation of aminocaproic acid in bacterial destruction of caprolactam, *Biochemistry Journal*, 1968, № 33, p.946 (in Rus.).

[52] Biodamage: Textbook. *Manual for biological faculties of universities*, Ed. by V.D. Ilyicheva, High School Publishing House, Moscow, 1987, 352 pp. (in Rus.).

[53] E.L. Pekhtasheva, *Topical issues of biological damage*, Nauka (Science) Publishing House, Moscow, 1983, 265 pp. (in Rus.).

[54] A.N. Neverov, *Actual problems of biological damage and protection of materials, components and structures*, Science, Scientific Council for biological damage of the USSR Publishing House, Moscow, 1989, 256 pp. (in Rus.).

[55] A.N. Neverov, *Biodamage and methods for assessing the biological stability of materials,* Scientific Council for biological damage of the USSR Publishing House, Moscow, 1988, 140 pp. (in Rus.).

[56] A.N. Neverov, *Biodamage, methods of protection*, The Scientific Council for biological damage of the USSR Publishing House, Poltava, 1985, 182 pp. (in Rus.).

[57] I.G. Kanev, *Biological damage of industrial materials*, Nauka (Science) Publishing House, Leningrad, 1984, 232 pp. (in Rus.).

[58] *Biological problems of environmental materials*, Ed. by A.N. Neverov, Scientific Council on the biological damage of Sciences Publishing House, Penza, 1995, 108 pp. (in Rus.).

[59] *Methods for determining the biological stability of materials*, Ed. by N.A. Plate, Science Council of biological damage of the USSR Publishing House, Moscow, 1979, 202 pp. (in Rus.).

[60] Microorganisms and lower plants - *the destroyers of materials and products,* Ed. by N.A. Plate, Nauka (Science) Publishing House, Moscow, 1979, 225 pp. (in Rus.).

In: Organic Chemistry, Biochemistry, Biotechnology ... ISBN: 978-1-62081-155-9
Editors: G. Zaikov, F. Pudel and G. Spychalski © 2013 Nova Science Publishers, Inc.

Chapter XIV

Biodamages and Protection of Paper

***Elena L. Pekhtasheva**, Anatoly N. Neverov, Irina G. Shishkina, and Gennady E. Zaikov*[1]**

G.V. Plekhanov Russian Economic University, Moscow, Russia
[1]N.M Emanuel Institute of Biochemical Physics, Russian Academy of Sciences, Moscow, Russia

Abstract

The article is collecting the information about biodamages of different kind of papers and discussing the ways for protection of papers against action of fungi. The contributors gave some recommendations for increasing of paper stability against actions of fungi.

Keywords: Biodamages, papers, fungi, protection, stability, degradation, deterioration

Paper is a material that largely consists of plant raw material processing products, primarily, cellulose, and contains sizing agents, mineral fillers, chemical and natural fibers, pigments and dyes. The mass of 1 m^2 of paper may reach 250 g. Paper with 1m^2 mass greater than 250 g is called cardboard in Russia.

The main raw material for paper and cardboard manufacture is coniferous and leaf wood. The plant raw material is processed into wood pulp and cellulose. Along these half-finished materials, the secondary raw materials (scrap paper and rag stock) are widely used for paper and cardboard production.

Paper and cardboard manufacture starts from wood pulp mixing in a particular proportion (composition). The mixture is pulped and milled in water using special equipment that converts it to the paper grade pulp.

* E-mail: pekhtashevael@mail.ru.

To make paper less absorbent and suitable for writing, sizing agents (plant and animal adhesives, soluble glass, synthetic substances, etc.) are added to the paper pulp. The surface structure, whiteness and other consumer properties are improved by adding fillers (kaolin, talcum, gypsum, chalk, etc.).

Paper pulp preparation is finished by its cleaning from mineral and organic raw particles. After that paper pulp is strictly defined concentration is delivered to moving machine wire. Paper grains precipitate on the machine wire and form the paper web, and water is strained out and drained.

Paper and cardboard quality is characterized by customer properties, which parameters are regulated by standards. The most important properties are: composition due to wood pulp type, 1 m^2 mass, thickness, density, smoothness, sizing degree, ash content, whiteness and dirtiness. Paper and cardboard are also characterized by strength at break, linear strain at moistening and drying, transparency, air permeability and other properties.

Depending on the purpose of application, paper is divided into several groups: printing paper; decorative paper; paper for writing, typewriting, charting and drawing; electrotechnical paper; packaging and wrapping paper; photosensitive paper; cigarette tissue; nonprinting paper for various purposes, etc.

Depending on the purpose, cardboard can be book board, packaging board, shoeboard, electrical pressboard, etc.

Given that paper is a material generally produced from natural cellulose which resources are quite considerable, it may compete with other natural polymers, as a nutritive substrate for microorganisms. Annually, huge masses of the customer paper are damaged and destroyed by microorganisms inducing the record loss.

Along with the material loss due to paper damaging by microorganisms, the inestimable, grievous loss of handwritings and hard copies important for science, art and history should be considered.

In this connection, analysis of various purpose paper and cardboard biodamages, the study of damage mechanism and types of microorganisms inducing it, determination of factors promoting and preventing development of this process, and searching for protective means against damages by microorganisms represent the urgent task, successfully solved in the modern science and practice.

In the paper making industry, biological growths are developed on processing equipment of factories, in water tanks and pipelines containing liquid wood pulp. Cellulose and pulp suspension, cooking liquor residues and wood extracts form favorable conditions for abundant microflora multiplication during the paper pulp manufacture. These growths are formed by bacteria and fungi and represent a loose mucous mass covering internal walls of the equipment. Buried in the growths are a fiber and a filler used in the paper making [1].

The growth accumulation rate equals 4-12 g of absolutely dry mass per 1 m^2 a day. When the mass reaches about 100-150 g per 1 m^2, growths fall down from the surface. Falling down in a mass flow during paper making, they lead to dirty spots and holes formation in the prepared paper. They also cause paper web breaks and associated decrease of equipment performance.

Paper microflora composition is in many instances provided by the process of its manufacture, storage and operation environment. Therefore, microorganisms inhabiting on the paper and inducing it damage shall be considered in connection with the production field and the surrounding.

In the paper making, carriers of microorganisms, which damage paper, can be: applied raw materials and equipment, water and air.

All initial raw materials feature the abundance of fungi and bacteria, the species composition being extremely manifold. Each period of paper making is also characterized by different species composition of microscopic fungi with respect of humidity and production regimes.

The specific feature of paper making microscopic fungi is low resistance to unfavorable factors, high temperature, in particular. Therefore, paper pulp microorganisms die during dewatering of the paper web in a drying machine at 120°C.

Thus fungi are virtually absent on the paper surface immediately after manufacture and then appear on it during packing, storage, transportation and making products from it, whereas bacteria steadily remain at all production stages. Of special importance are spore-forming bacteria *Bac. subtilis.*

Cellulose and pulp suspension, cooking liquor residues and wood extracts form favorable conditions for microorganism development during paper pulp preparation. In this case, growths are mostly formed by bacteria and, to a lesser extent, by fungi. Among bacteria, *Bacillus* genus dominates, e.g. thermophilic variants *Bac. Subtilis* with optimal multiplication temperature of +45…+50°C. Meanwhile, it is noted that growth of some microorganisms may be intensified by the effect of others. For example, bacteria suppress growth of some fungi, whereas some fungus species stimulate growth of one another.

Microscopic fungi mostly colonize paper during storage. Multiple cases of printing product damaging by microscopic fungi in the course of both production (e.g. specific microflora typical of bindery conditions) and book storage in shopping spaces are described [18].

The quantity of microorganisms on the surface of paper products increases quickly due to their use and environmental hygienic conditions. For instance, books returned by reader to libraries show four-fold greater amount of fungal spores than newly delivered paper products.

The species composition of microflora on paper also depends on materials contacting with it during storage: chemical fibers, real, artificial and synthetic leather, lacquers, paints, various plastics and binders.

Biostability of paper depends on many factors: the type and method of fiber processing, quality and composition of fillers and sizing agents, production technology features, and operation conditions. The basic condition of paper preservation against microorganisms' impact is air humidity maintaining below 55%.

Among the whole diversity of microorganisms inhabiting on paper, the main degraders of it are fungi and cellulose-fermenting bacteria of various species and genus. The list of fungi able to multiply on paper includes 308 species of 132 genus now.

Fungi can be divided into five groups with respect to the damage rate and size caused to paper products [1]:

- Fungi always observed on paper which cause its complete decay (*Afumigatus, A. terreus, etc., totally, 25 species);*
- Fungi always observed on paper which cause just some deterioration of its texture (*A. niger, A. flavus*, etc.);

- Fungi consuming some nonspecific components of paper – wax, paraffin, asphalt, wool, rubber, synthetic polymers, etc. The composition of this group is variable and expands, as new compounds in the paper composition are used (*Oidium dioxamii*, etc.);
- Fungi which presence on paper depends on the surrounding microflora. This group features predominance over the main paper degraders;
- Occasional representatives of fungi.

Various sorts of paper infected by Penicillium ochrochloron fungus (frequently determined in books) possess different biostability (Table 1) estimated by mechanical strength loss, which is determined by folding endurance.

The fungal effect on paper is a complex of chemical and biochemical processes, in the course of which paper damage is determined not only by its consumption as a nutritive substrate that depends on its physical features as a capillary-pore system, in particular. Inhomogeneous capillary-pore systems with discrete foci, favorable for damaged initiation are formed in paper. Paper capillaries are filled with air and water. The latter, in many instances, defines paper properties, including biostability. When air humidity is 30-40%, capillary water is almost absent in cellulose and microorganisms cannot multiply under this condition. At 55-65%, the content of capillary and bound water is nearly the same that gives somewhat a transition zone. And at 70-75%, condensed water is observed in quite large capillaries what allows microorganisms for germinating and multiplying on paper.

Cardboard contains heterogeneous materials in its structure, including porous components which easily accumulate water. Water is one of the basic factors which determine manifestation of physiological functioning of micromycetes on the cardboard. In its presence, fungal spores occasionally appeared on the cardboard begin germinating.

Table 1. Funginertness of different sorts of paper

Sort of paper	Funginertness	Mass loss, %	Spot color in damaged areas	Damaged area pH
Cotton	Low	-	Brown	Slightly increases
Enamelled	High	5	Bright yellow	Decreases
Sulfite	Medium	2.3	From maize yellow, brown to violet	Increases
Filter	Low	5.2	Yellow, orange, brown and green spore bloom	Slightly decreases
Newsprint	Very low	5.5	Pale orange, dark green dots	Slightly decreases
Book and magazine	Low	3.9	Maize yellow and orange, spore bloom	Slightly decreases

Growth and development of fungi gradually intensify, involving the surface and deeper layers of the material. Fungal mycelium is capable of accumulating water, and materials even with low water content become accessible for fungi. Water is also synthesized as a result of physiological and biochemical activity of fungi. Cardboard converts to a colloid-biological system, in which capillaries are frequently filled with mycelium of fungi excreting various metabolites, including enzymes, into the environment.

Normal development of fungi requires definite amount of oxygen. Therefore, cardboard increased porosity and permeability to air are advantageous for it. As shown, the decisive factor for fungus development on the cardboard is capillary water content that fills fiber pores rather than chemical composition and physicochemical properties of the substrate. Cardboard destruction intensity is most significantly intensified in the presence of easily degradable products, mostly various binders, covers, leather, etc.

Cardboard damaging by fungi is closely associated with their ability to synthesize enzymes, cellulose-fermenting ones, first of all. Cardboard cellulose is highly hydrated at increased air humidity that, in turn, significantly increases damageability of the cardboard. Cardboard aging and degradation are accompanied by synthesis of low-molecular products and generally reduces its microbiological stability, because using these easily assimilable substances fungi start intensive functioning and excreting a variety metabolites into the environment that accelerates degradation of the cardboard fibers.

Microorganisms diversely lower quality of paper. Microscopic fungi which develop exclusively on the paper surface stain it differently. For instance, *Pullularia pullulans* fungus makes paper light blue, *Trichoderma* or *Penicillium* - green, and *Cladosporium* - dark.

Damaging cellulose fibers of paper, molds destroy them from the surface and perforate paper, growing into it.

High concentration of *Aspergillus* genus representatives on the cardboard should be outlined. These species easier endure thermal treatment and sometimes, their spores survive even in the course of cardboard production procedure. Thus, there is potential possibility of cardboard damage by *Aspergillus* genus fungi immediately after manufacture [2].

Not only books, documents, maps and other paper products for long-term storage require protection against microorganism damage. Paper and cardboard for packaging, especially of food and goods deliverable to tropics, must also be highly biostable.

Paper used for external insulation in cable production; paper and cardboard for packing soap; paper used for storage of various ammunition; and special paper preserving goods from moth impacts shall also be high resistant to microorganisms' action.

Development of trade, cultural, political and economic relations with foreign countries located in the regions with hot and humid climate necessitates creation of special sorts of paper and cardboard both resistant to fungal and bacterial damages and manifesting antimicrobial action. This enables their use for protection of some goods against damaging during their storage and operation. For example, this shall be the feature of the cigarette type tissue paper used for packing ferrous and non-ferrous metal products and their protection against corrosion, box board for bearings delivered to foreign countries, etc. [3-15].

In the paper-making industry, over 300 compounds were specially synthesized as protective means against biological growing. However, just few of them have fount practical application, because biocides for paper should have the following features [16]:

- Be water-soluble or form stable aqueous emulsions;
- Do not change properties at temperature up to +70°C and in pH 4-7 range;
- Not be sorbed by cellulose fibers and not form chemical bonds with them;
- Have high selective toxicity to oxidizing microflora of biological treatment plants.

The properties of applied biocides must be stable and durable. Selection of such compounds is a challenging task. There is no preparation until now, which would comply with the imposed requirements.

The following compounds are used as biocides with sufficiently high protective properties in Russia: salicylanilide and various halogen derivatives, oxyquinoline derivatives (sodium salt, etc.), some disulfides, sodium pentachlorophenolate, copper naphthenate, etc.

For example, the use of sodium pentachlorophenolate for microbial growth suppression at a plant producing newsprinting paper has increased equipment performance, decreased paper breakage rate and enabled considerable savings. In this case, elimination of microbial impact represents a high reserve for increasing paper production volume even on currently existing equipment.

The following methods are used for injecting biocides into paper: paper and cardboard impregnation; antiseptic application on the paper surface; paper impregnation by dispersion containing biocide and hydrophobic additives; biocide injection to fibrous suspension.

Gaseous biocides of oxyethylene and formaldehyde type are used to eliminate microflora damaging archive documents.

The use of polymeric materials, e.g. poly-*para*-benzene dicarbinol coating, is one of the direction in the field of biostable paper production. This coating is used for preservation of documents and not only increases paper strength, but also protects it against damaging by microorganisms.

Paper treatment by polymers with hydrophobic properties, polyethylene, for example, forms protective barrier much more effectively than the treatment by fungicides.

Paper and books protective measures against biodamages include air conditioning to maintain the optimal temperature and humidity regime, air cleaning from dust and air filtration and sterilization by UV-light, if required, as well application of chemical protection means – biocides for surface treatment or preparation of treated paper.

References

[1] J.P. Nyuksha, *Biological damage to paper and books*, Nauka (Science) Publishing House, St.-Petersburg, 1994, 235 pp. (in Rus.).

[2] S.A. Vilkova, *Commodity Expertise and toiletries*, Business Books Publishing House, Moscow, 2000, 286 pp. (in Rus).

[3] E.L. Pekhtasheva, *Topical issues of biological damage*, Nauka (Science) Publishing House, Moscow, 1983, 265 pp. (in Rus.).

[4] A.N. Neverov, *Actual problems of biological damage and protection of materials, components and structures*, Science, Scientific Council for biological damage of the USSR Publishing House, Moscow, 1989, 256 pp. (in Rus.).

[5] A.N. Neverov, *Biodamage and methods for assessing the biological stability of materials*, Scientific Council for biological damage of the USSR Publishing House, Moscow, 1988, 140 pp. (in Rus.).

[6] A.N. Neverov, *Biodamage, methods of protection*, The Scientific Council for biological damage of the USSR Publishing House, Poltava, 1985, 182 pp. (in Rus.).

[7] I.G. Kanev, *Biological damage of industrial materials*, Nauka (Science) Publishing House, Leningrad, 1984, 232 pp. (in Rus.).
[8] *Biological problems of environmental materials*, Ed. by A.N. Neverov, Scientific Council on the biological damage of Sciences Publishing House, Penza, 1995, 108 pp. (in Rus.).
[9] *Methods for determining the biological stability of materials*, Ed. by N.A. Plate, Science Council of biological damage of the USSR Publishing House, Moscow, 1979, 202 pp. (in Rus.).
[10] Microorganisms and lower plants - *the destroyers of materials and products,* Ed. by N.A. Plate, Nauka (Science) Publishing House, Moscow, 1979, 225 pp. (in Rus.).
[11] *First Conference on Biological Damage*, Ed. by N.A. Plate, Nauka (Science) Publishing House, Moscow, 1978, 226 pp. (in Rus.).
[12] Problems of biological damage to materials. *Environmental aspects* Ed. by A.N. Neverov, Scientific Council of biological damage of the USSR Publishing House, Moscow, 1988, 124 pp. (in Rus.).
[13] *Environmental problems of biodegradation of industrial, construction materials and industrial wastes*, Ed. by A.N. Neverov, Scientific Council on Problems of biological damage Publishing House, Penza, 2000, 192 pp. (in Rus.).
[14] *Fourth Conference on Biological Damage*, Ed. by N.A. Plate, Scientific Council of biological damage of the USSR Publishing House, Nizhnii Novgorod-city, 1991, 99 pp. (in Rus.).
[15] Biodamage: Textbook. *Manual for biological faculties of universities,* Ed. by V.D. Ilyicheva, High School Publishing House, Moscow, 1987, 352 pp. (in Rus.).
[16] K.Z. Gumargalieva, G.E. Zaikov "*Biodegradation and biodeterioration of polymers. Kinetical aspects* ", Nova Science Publ., New York, 1998, 210 pp.

In: Organic Chemistry, Biochemistry, Biotechnology ... ISBN: 978-1-62081-155-9
Editors: G. Zaikov, F. Pudel and G. Spychalski © 2013 Nova Science Publishers, Inc.

Chapter XV

Biodegradation and Protection of Synthetic Polymer Materials

***Elena L. Pekhtasheva**, *Anatoly N. Neverov*, *Irina G. Shishkina and Gennady E. Zaikov*#**

G.V. Plekhanov Russian Economic University, Moscow, Russia
#N.M Emanuel Institute of Biochemical Physics,
Russian Academy of Sciences, Moscow, Russia

Abstract

The paper is including information about biodegradation and biodeterioration of plastics and materials on the base of plastics by different bacteria: Penicillium, Aspergillus, Chaetomium, Fusarium, Alternaria, Trichoderma, Rhizopus *and genus fungi. The contributors are discussing the problems of biostability for* polyethylene, polypropylene, polystyrene, polyvinyl chloride (rigid), polyamide, and polyethylene terephthalate and materials on the base of these polymers.

Keywords: Biodegradation, protection, synthetic polymers, plastics, materials, biodamages

1. Biodegradation and Protection of Plastics

Materials based on polymers, which are in the viscoelastic or rubbery state at the molding stage and solid during operation are called plastics.

Polymers are high-molecular compounds consisting of long chain molecules formed by multiple alterations of equal groups of atoms, linked with one another by chemical bonds.

* E-mail: pekhtashevael@mail.ru.

Along with polymers, plastics comprise fillers, plasticizers, dyes, stabilizers and other additives.

Synthetic polymer materials are widely used in virtually all fields of science and technology, in industry, construction, agriculture, etc.

Synthetic polymers are more resistant to degradation by microorganisms than natural high-molecular compounds. Polymeric backbone of the macromolecule of synthetic high-molecular compounds is too large and strong to be directly consumed by bacteria or fungi. However, in some instances, even they are degraded by microorganisms [1-51].

Insects and rodents can also destroy polymers. Biological damages of plastics by insects and rodents are manifested in direct mechanical destruction of some parts, protective coatings and packing materials.

Destruction of plastic package and further insect and rodent colonization and propagation may happen in components of instruments and mechanisms, difficult to access for a man, which may be a safe ecological niche for animals. For many times, accumulation of animals and their metabolites in heavy duty parts of electric appliances caused short circuits and other failures.

Most frequently, damages are caused by Penicillium, Aspergillus, Chaetomium, Fusarium, Alternaria, Trichoderma, Rhizopus, *etc. genus fungi.*

Mold fungi induce chemical (by metabolites) and mechanical (growing, mycelium hypha penetration into the material) damages of the materials. Extracellular enzymes and organic acids are basic chemical products of fungus metabolism, which induce damages of synthetic polymer materials via chemical degradation (hydrolysis, oxidation, etc.) of polymer macromolecules or low-molecular components (fillers, plasticizers, etc.).

Along with purely chemical degradation of polymeric materials, microorganism and metabolites can cause changes of their physicochemical and electrophysical properties, as a result of swelling and cracking. Deterioration of decorative and other exterior properties of polymeric materials resulting biogrowth (mold spots occurrence, although in this instance performance of the article can be preserved) is possible.

The development of mold fungus culture of the polymer surface promotes water vapor condensation from the atmosphere, water accumulation, and only this circumstance can have an adverse effect on the change of polymer material properties. Chemical interaction of microorganism metabolites with a polymer or additives of synthetic material may result in changes of some physicochemical properties of the material. Properties of fungus unstable materials, such as strength, flexibility, dielectric parameters may decrease, insulating properties may deteriorate, color of stained surfaces may change, etc.

Bacteria damage plastics less frequently, but their action may be insidious. In some cases, their presence can be hardly seen by the naked eye. The damage may be judged about by occurrence of foreign odor, staining, mucous, etc.

Various species and genus bacteria (Pseudomonas, Bacillus, etc.) participate in biodegradation of plastics.

Bacteria adapt to synthetic polymers and degrade high-molecular compounds of different chemical structure to low-molecular fractions using various enzymes and metabolites [52-64].

Along with other materials, biodegradation of plastics usually happen simultaneously with their aging under the action of external physical and chemical factors of the environment (UV-radiation, water, temperature drops, etc.). Biodegradation and aging, both these processes complement and aggravate each other.

Degradation of plastics depends not only on the species and genus of impacting microorganisms. The following parameters affect the degradation rate of plastics: chemical structure of the polymer itself, its physical structure, molecular mass, molecular-mass fraction distribution, the presence and composition of plasticizers, fillers, stabilizers and other additives.

There is a definite dependence between biodemage rate and chemical structure of the polymer. Biostability depends on the chemical origin, molecular mass and supermolecular structure of the polymer [1-51].

Types of bonds R - C_3 and R - CH_2 - R' are inaccessible or hardly accessible for microorganisms. In turn, unsaturated valences R = CH_2 and R = CH - R', as well as bonds of polymers most sensitive to hydrolysis, such as acetal, amide, ether and carbonylic or carboxylic ones, are easily accessible bonds to be broken by microorganisms.

The macromolecule size is the important factor defining polymer resistance to biodegradation. Polymers with high molecular mass are harder degradable by microorganisms, whereas monomers or oligomers can be easily degraded by microorganisms.

Supermolecular structure of synthetic polymers is equally important factor affecting biodegradation. Compact dislocation of crystalline polymer structure fragments restricts their swelling in water and, simultaneously, prevents enzyme penetration into their structure. Hence, the effect of enzymes on both polymer backbone and biodegradable parts of macromolecule chain are restricted.

The presence of defects in macro- and microstructure and molecular heterogeneity promotes biodegradation.

Biodamages of the Main Components of Plastics

Plastics are based on polymeric binders, which are polymeric resins. By type of polymeric resins, plastics are divided into thermoreactive and thermoplastic (with respect to the method of their hardening during production of the material), polyethylene, polyvinyl chloride, polyamide, etc. (with respect to chemical structure of the polymer) [52].

Carbochain polymers with the macromolecule backbone composed exclusively of carbon atoms (polyethylene, polypropylene, polyvinyl chloride, etc.) and heterochain polymers, which backbone, along with carbon, comprises oxygen, nitrogen and other atoms (polyamides, polyurethanes, etc.), are distinguished.

Polymeric resins demonstrate different biostability, depending on chemical structure of macromolecule, polymeric chain length, side branches, etc. The general rule is increasing resistance of polymers to microbiological damage with the macromolecule chain length. All other conditions being equal, linear carbochain polymers are less bioresistant than branched or heterochain ones.

The influence of chemical structure on biostability of polymers was determined on the example of polyurethanes. For this purpose, over 109 specimens containing no admixtures, on which microscopic fungi might develop, were synthesized. It is found that polyurethanes with ether bond were damaged stronger than these with ester bond. The presence of ether bond simplifies degradation and consumption of polymer. It is also found that compounds are also decayed, which have a long carbon chain between ether bonds. The presence of three neighboring methyl groups also increases damaging of polyurethanes by microscopic fungi.

It is found, for example, that microbiological stability of polymeric resins directly depends on molecular mass of the polymer and decreases in the presence of low-molecular fragments in the material. The same effect is observed in case of aging caused by light and heat.

Amorphous-crystalline phase transition of the polymer increases its biostability.

Among polymeric resins with increased resistance to damage by mold fungi, the following should be mentioned: polyethylene, polypropylene, polystyrene, polyvinyl chloride (rigid), polyamide, and polyethylene terephthalate. Less fungus resistant polymers are polyvinyl acetate, polyvinyl alcohol, chlorosulfonated polyethylene, etc.

Plasticizers are important components of plastics. They are mostly dicarboxylic and polycarboxylic aliphatic and aromatic acid esters. The plasticizer content may reach 30-50% of the plastic mass; therefore, to a great extent, biostability of the material depends on its biostability.

A dependence between biostability of organic plasticizers and the length and spatial configuration of the backbone is found: orthophthalic acid esters are most stable, whereas para-, metha-, iso- terephthalic acid derivatives possess the lowest stability.

Plasticizers of ether type are hydrolyzed to short-chain bases and acids, which are utilized by microorganisms. This process may proceed at relatively low air humidity (50%) and temperature (+20°C).

Consuming plasticizers and fillers as the nutrition source, microorganisms intensify aging of plastics.

The comparison of resistance to degradation by mold fungi of the most widespread plasticizers (phthalic and adipic acid ethers) indicates the highest stability of aromatic phthalic acid ethers rather than adipic acid ethers – aliphatic dicarboxylic acid. Aliphatic sebacinic acid ethers have low fungus resistance.

Fillers are important components of plastics. They generally represent inert solids injected to polymeric materials for the purpose of regulating mechanical and other properties. Injection of the filler also reduces the cost of materials and articles from plastics, increases their strength, electrical and other properties.

Organic fillers (wood powder, cotton fibers, paper, etc.) are nutritive substrates for microorganisms and reduce fungal resistance of polymeric composites, whereas inorganic fillers (asbestos, glass fiber, quartz powder, kaolin) increase biostability.

However, if the binder impregnates organic filler well and is highly resistant to molds, fungal resistance of the polymeric material can be quite high. It is important to achieve maximum impregnation of the filler by resin in the technological process. Sometimes, this is achieved under vacuum. Mechanical processing of finished plastic products with non-bioresistant fillers without appropriate protection of processed areas by stable to fungi lacquer coating is unallowable. Dynamics of plastics damage depends on both chemical composition and physical structure.

Fungus mycelium can propagate in extremely thin cracks and pores, formed at the interphase and surface boundaries in the material. For instance, ethylene-vinyl-acetate copolymers were damaged with fungus hyphae development at the boundary of the polymer and starch grains. Meanwhile, the damage increased with vinyl acetate content in the material and was stimulated by adding starch, as the filler.

Polyethylene is a carbochain thermoplastic polymer, one of the most widely applied polymers of the polyolefin sequence. It is used for manufacture of films, protective coatings, insulating articles, containers, package materials, etc. It has high dielectric properties and chemical stability.

Microbiological resistance of polyethylene is characterized by a property, general for all alkanes – the higher molecular mass, the higher biostability of the material. Polyethylene is usually damages from the surface, most degradable being the polymer with the molecular mass below 25000. Biostability of high density polyethylene is higher than biostability of low density polyethylene.

When exposed to soil in moderate climate, polyethylene products are assumed 8-year resistant to microbiological damages. In tropics, the effective life decreases.

Polyethylene surface overgrown by mold becomes rough and covered with mosaic black-brown spots.

Polystyrene is a carbochain thermoplastic polymer obtained by styrene polymerization in the presence of various initiators. It is water resistant and has high dielectric properties. Radio and electric equipment parts, insulating films, foamed plastic, etc. are produced from polystyrene.

Polystyrene is resistant to microorganisms: 8 month exposure to a mixture of molds causes not damage.

Polyamides are heterochain thermoplastic polymers with - CO - NH – group in the backbone. They possess increased mechanical properties, good dielectric properties; however, they are low resistant to light and oxidant impacts.

Polyamides are used in production of films, fibers and various products.

Note that the in-depth study of biostable polyamide materials was preceded by quite detailed investigation of ε-caprolactam, which is the primary material for ε-polycaproamide synthesis, the latter being the most widespread kinds of fibrous polyamide materials.

In this case, the study of caprolactam degradation processes caused by microorganisms should be considered as a model for studies of microbial degradation of polycaproamide materials.

It is found [65-75] that during biochemical sewage treatment caprolactam is completely degraded by microorganisms of active sewage silt.

Among 30 strains, only 4 species from *Eubacteriacede* family and *Bacillus*, *Bacterium* and *Pseudomonas* genus were found capable of propagating on a mineral medium containing caprolactam.

Japanese scientists have extracted pure bacterium cultures of *Pseudomonas, Corynebacterium* and *Achromobacter* genus able to consume caprolactam as the unique source of carbon and nitrogen.

It is author's opinion [65-75] that *Micrococcus varians* strain extracted from the soil is able to destroy caprolactam.

Complete hydrolysis of ε-caprolactam in concentration as high as 3 g/l by spore-forming bacteria from *Bacillus subtilis- mesentericus* group at their continuous cultivation proceeds within 24 hours.

Note also that caprolactam consumption by bacteria is clearly adaptive: when grown on nutritious media, bacteria quickly become unable to oxidize caprolactam.

Some microorganisms use caprolactam as the source of carbon and nitrogen, others require an additional source of carbon and use caprolactam as the carbon source exclusively.

A scheme of ε-caprolactam metabolism due to bacterial degradation is suggested [65-75]. This scheme includes ε-caprolactam hydrolysis to ε-aminocaproic acid by lactamase enzyme with further synthesis of L-ketoglutaric acid, which then is transaminated by transaminase enzyme to glutamic acid. Meanwhile, adipic acid half aldehyde is synthesized, which is then converted by dehydrogenase to adipic acid.

It is found that the general scheme of ε-caprolactam metabolism caused by bacteria is analogous to the one previously suggested by Japanese scientists. The first stage of caprolactam degradation is hydrolysis by amide bond to ε-aminocaproic acid. Similar to caprolactam, aminocaproic acid is far not normal cell metabolite. Nevertheless, its further transformation proceeds by way common for natural amino acids. Significant ammonia quantities in the culture fluid were detected. Ammonia excretion is associated with deamination of ε-aminocaproic acid. As a result of ε-caprolactam decay, volatile acetic and butyric acids and neutral products (aldehydes, ketones and alcohols) are formed.

In the *Achromobacter guttatus* culture medium degrading caprolactam, glutamic acid, lysine, threonine and methionine are identified.

Laboratory examination for microbiological stability of film materials (polyamide-6, polyamide-6,6) on optimal nutritive medium indicates that these materials are bio-unstable. All samples demonstrate surface and through destruction of polymers. All samples demonstrate surface and through degradation of polymers. Polyamide-12 demonstrated high resistance to molds. Impact of some strains of fungi on polyamide films (PK-4 film) reduces strength to 80%.

Polyvinyl chloride is a carbochain thermoplastic polymer, one of the most widely applied to manufacture of rigid and plasticized polymers for pipe, sheet, film, shaped profile, fiber, protective coating, etc. production.

Polyvinyl chloride resin is combined well with many plasticizers. Flexible PVC is the base for artificial leather production, widely used in aircraft and automobile industry, agricultural engineering, personal protection equipment production, production of shoes, light industry products, etc. Biostability of plasticized polyvinyl chloride significantly depends on biostability of applied plasticizers, stabilizers, etc. Plasticized polyvinyl chloride film loses strength, resulting microorganism impact, which is also accompanied by mass reduction and rigidity increase. Different film staining (red, orange, pink) is observed and light transmittance is reduced.

A large body of research point at connection of observable changes withinsufficient biostability of plasticizers and the rate of their migration from the volume of flexible PVC.

Rigid polyvinyl chloride is more biostable in relation to bacteria and molds. After 8-year tests in the soil, pipes from rigid polyvinyl chloride have not considerably decreased their physicomechanical properties, although the material itself is able to support fungus growth.

Polycarbonates are heterochain thermoplastic polymers synthesized by polycondensation of carbonic acid ethers and dioxy-compounds.

Polycarbonates possess increased mechanical properties, water and weather resistance, are good dielectrics and physiologically innocent. They are used in production of electronics, watches, refrigerators, films, dishware, etc.

The study of microbiological resistance of one of polycarbonates has shown that molds are able to propagate on the material surface at 100% air humidity and temperature of +30°C. Polycarbonate film is also not exactly fungus-resistant. This should be considered at its application in electrical appliances, for medical purposes and packaging. Polycarbonates are resistant to the impact of bacteria.

Polyurethanes are heterochain thermoplastic polymers with urethane group in the macromolecule. It is characterized by multiple applications in technology. Flexible and rigid foamed plastics, elastomers, fibers, films, adhesives, lacquers, solid and flexible covers are produced from polyurethanes.

As compared with polyolefins, polyurethane-based synthetic polymeric materials possess lower resistance to fungal action.

Comparative study of microbiological stability of polyurethane polymers synthesized from polyethers and polyesters shows that polymers from polyesters are more damageable by mold fungi than polymers from polyethers.

Polyurethane protective coats of aviation aluminum fuel tanks and metal structures are highly damaged by fungi down to exfoliation and full degradation.

Along with the structure and composition of plastics, their biostability is significantly affected by environmental conditions: high air humidity, increased temperature, difference in day and night temperatures. Water vapor condensation and moisture accumulation on the surface material promote propagation of microorganisms. Some plastics, already affected just by high water content, change their properties. This effect is added by chemical corrosion induced by microorganism metabolites, which results in deterioration of properties and quality of the products.

Microorganisms of various groups deteriorate mechanical, hygienic and esthetic properties of plastics.

Pigments from microorganisms stain plastics: grey, green, violet or pink spots occur; bleaching or pitting of the surface is also possible.

Sometimes, damages happen on the surface and are only manifested in mycelium growth, which can then be removed and, consequently, will cause no considerable effect on entire material or article performance. In other cases, biodamages may be deeper, when along with the appearance, physicochemical, physicomechanical and other properties of materials change, e.g. the change of viscosity, strength, hardness, insulating and other properties is observed.

In connection with plastic articles use under conditions suggesting high impact of microorganisms, the problem of polymeric material protection is extremely urgent.

At present, over 3000 compounds possessing biocidal properties are described. However, no antiseptics complying with the requirement to them are found yet.

Despite the fact that the majority of known biocides are probed on plastics, they have no practical application. This is due to specificity of plastics manufacture. When manufactured and processed, plastics are subject to impact of high temperatures allowable just for few biocides.

Along with thermal stability, biocides shall be chemically stable: this means that biocides must not interact with other components of plastics and, at the same time, be compatible with the plastic (the polymer and all its components).

One of the requirements imposed on biocides for plastics is a wide spectrum of antimicrobial action at low concentrations, because high biocide concentrations may decrease physicomechanical and electrical properties of articles from plastics.

Moreover, biocide must be harmless and shall not be washed off during operation. Since a number of specific requirements are imposed on the articles from plastics, e.g. electric resistance and dielectric properties, biocides must be non-polar compounds.

All these requirements limit the number of biocides used for plastics protection. Among biocides for plastics, during the years the following substances are used: salicyl anilide, copper 8-oxyquinolate, 2-oxydiphenyl, 4-nitrophenol, sodium pentachlorophenolate, etc. The following biocides are also known: trilan (4,5-trichlorobenzoxazolinone), cymid (dichloromaleic cyclohexylimide), and some organoarsenic and organotin compounds.

Trilan, cymid and epoxar (an arsenic-containing compound) have shown good results as biocides for fungus-resistant films from polyvinyl chloride and artificial leather for technical purposes used in articles, which are to be delivered to tropics. For example, addition of 1-2% of cymid to artivicial leather films made of polyvinyl chloride provides long-term preservation of strength, good appearance and other properties of the material under the most severe conditions. The strength of material not protected by trilan, under the same conditions, is reduced by 15-30% within three months.

The advantage of epoxar is that along with biocidal properties, it is able to improve light and thermal stability of polymeric materials, i.e. is the universal stabilizer.

Sometimes, fungicidal treated polymeric materials allow a trick for complex technical issues. Films containing biocides were successfully used for lining bottoms of channels, pools and other hydraulic works. Such protective coatings were not overgrown by microorganisms and algae, had increased lifetime, prevented water leakage to soil, and even prevented, to a definite extent, water against "contamination" by microorganisms and algae.

The use of biocides in plastics can be intended for both preventing them against biodegradation and hygienic purposes. For example, in some medical institution biocidal plastics have proved themselves in production of plastic handles, water-closet seats, some parts of medical equipment, film articles, antiseptic linings for bassinets, etc.

Hygienic purpose articles made from polyethylene, shock-resistant polystyrene and other plastics fungicidal treated by hexachlorophene (1-2%) retain antiseptic properties after a year of operation in hospitals and prevent contagion, whereas under the same conditions, common, biocide unprotected materials can be the infection foci.

2. Biodegradable Polymeric Materials

Waste polymeric materials, extremely slow degradable under natural conditions, are considerable source of environmental pollution.

Single-use plastic package, films and packing materials are of special hazard. That is why, along with solving the problem of increasing plastics durability, special types of polymers with regulated lifetime are designed and manufactured. The adverse feature of these polymers is their ability to retain consumer properties during the entire operation period and only then be subject to biological and physicochemical transformations.

Owing to chemical structure and high molecular mass, the most of volume plastics types (polyethylene, polypropylene, polystyrene, PVC) degrade very slowly. Soil microorganisms might degrade these polymers, if their long macromolecules would split into shorter fragments [1-51].

There are several approaches to creation of biodegradable polymers:

- Creation of photodegradable polymers, which are able to degrade under natural conditions to low-molecular fractions due to special additives present in them. Further on, these fractions are degraded by soil microflora;
- Designing of polymeric composites containing, along with high-molecular backbone, organic fillers, which are the nutrition for microorganisms (starch, cellulose, pectin, amylase, etc.);
- Creation of polymers with the structure similar to natural polymers;
- Polymers can be synthesized by biotechnology methods.

One of directions in creation of photodegradable polymers is inclusion of chromophores into polymeric backbone, which provide UV-light absorption by polymers that induces their degradation.

At present, studies associated with development of composites containing, along with high-molecular backbone, organic fillers, which are the nutritive medium for microorganisms are of interest. Meanwhile, beside material degradation associated with filler destruction by bacteria, an effect of additional degradation stipulated by the features of filled polymer structure is observed. It is known that the filler may be accumulated in lesser ordered polymer zones. Moreover, macromolecule packing density in the interface layer of the "polymer-filler" system is approximately twice lower, than in the rest volume of disordered phase of the polymer. Therefore, when the filler is destroyed by bacteria, microorganism access to less stable in relation to biodegradation of a part of polymer becomes simpler [52].

Curious Facts

American Warner - Lambert Company has designed the first polymeric material Novolon, which only consists of starch and water and is completely biodegradable. This polymer can be processed by traditional methods, and mechanically, is intermediate between polystyrene and polyethylene.

Archer Daniels Midland Company (USA) has designed a polyethylene-based concentrate of Polyclean trademark used for production of biodegradable films. The concentrate contains 40% of starch and an oxidizing additive: starch quantity in the final product is 5-6%. The oxidizing component acts as the catalyst both in the light and in darkness. Starch degradation makes access of microorganisms and oxygen to polymer surface simpler, i.e. a specific synergic effect is observed.

ICI Americas Inc. company produces a heat-sensitive plastic, which is naturally degraded. It has properties similar to polypropylene. This is linear polyether (poly-3-hydroxybutyram-3-hydroxyvalerate), which is produced by sugar fermentation with the help of Alcfligenes eutrophys bacteria. This material is degraded by microorganisms, inhabiting in the soil, sewage and at water body beds.

Of great interest are studies aimed at creation of materials, which are not only biodegradable, but are also produced from recoverable biological resource, given that traditional raw material sources for the polymer synthesis are limited. These polymers are mostly based on starch, products of bacterial fermentation of sugar, heat-sensitive plastics based on animal starch with added petrochemical products.

Polymers may also be turned biodegradable by breeding specific microorganism strains able to degrade polymers. This direction is only successful in relation to polyvinyl alcohol yet. Japanese scientists have extracted Pseudomonas bacteria from the soil, which produce an enzyme cleaving polyvinyl alcohol. After degradation, polymer fragments are completely consumed by these bacteria. Therefore, Pseudomonas bacteria are added to activated sewage sludge for more complete cleaning from this polymer.

It is found [68] that regularly alternating co-polyamides containing L-amino caproic acid as one of the components are easily biodegradable, and glycine and L-amino caproic acid co-polyamide, degraded by bacteria and fungi in several weeks have been synthesized.

Polyamide synthesized from benzyl malonic acid and hexamethylene diamine is highly biodegradable. Such polymers [68] may be used as hypoallergic surgical threads, capsules for seedlings at recovery of forests, nonwoven materials and films used at mature tree moving as a single-use package. Other fields of application are also suggested: fishery, protection against floods, etc.

References

[1] M.N. Rotmistrov, P.I. Gvozdyak, S.S. Stavskaya, *Microbial degradation of synthetic organic compounds,* Naukova Dumka Publishing House, Kiev, 1975, 224 pp. (in Rus.).

[2] R. Blagnik, B. Newly, *Microbiological corrosion*, Khimiya (Chemistry) Publishing House, Leningrad, 1965, 222 pp. (in Rus.).

[3] I.A. Yermilov, *Theoretical and practical bases of microbial degradation of chemical fibers,* Nauka (Science) Publishing House, Moscow, 1991, 248 pp. (in Rus.).

[4] L.A. Wolfe, A.I. Meos, *Fibers for special applications*, Khimiya Chemistry) Publishing House, Moscow, 1971, 224 pp. (in Rus.).

[5] V.S. Bugorkov, T.A. Ageev, V.M. Galperin, The basic directions for creating photo-polymer materials and biodestruktiruemyh, *Plastics Journal,* 1991, N 9, p. 48-51 (in Rus.).

[6] O.A. Legonkova, L.A. Sukharev, P*olymer biological resistance to biodegradation,* RadioSoft Publishing House, Moscow, 2004, 272 pp. (in Rus.).

[7] K.E. Perepelkin, *The present and future man-made fibers*, LegPromBuzines Publishing House, Moscow, 2000, № 8 (22), № 9 (23), № 10 (24), № 11 (25), № 12 (26), (in Rus.).

[8] N.M. Emanuel, A.L. Buchachenko "*Chemical physics of degradation and stabilization of polymers*", VSP International Science Publ., Utrecht, 1982, 354 pp.

[9] A.F. Bochkov, G.E. Zaikov "*Chemistry of the glycosidic bonds. Formation and cleavage* ", Pergamon Press, Oxford, 1979, 210 pp.

[10] S.D. Razumovskii, G.E. Zaikov "*Ozone and its reactions with organic compounds*", Elsevier, Amsterdam, 1984, 404 pp.

[11] N.M. Emanuel, G.E. Zaikov, Z.K. Maizus "*Oxidation of organic compounds. Medium effects in radical reactions* ", Pergamon Press, Oxford,1984, 628 pp.

[12] V.A. Afanasiev, G.E. Zaikov "*In the realm of catalysis*", Mir Publishers, Moscow, 1979, 220 pp.
[13] Yu. V. Moiseev, G.E. Zaikov "*Chemical resistance of polymers in reactive media*", Plenum Press, New York, 1987, 586 pp.
[14] G.E. Zaikov, A.L. Iordanskii, V.S. Markin "*Diffusion of electrolytes in polymers*", VNU Science Press, Utrecht, 1988, 328 pp.
[15] K.S. Minsker, S.V. Kolesov, G.E. Zaikov "*Degradation and stabilization of polymers on the base of vinylchloride*", Pergamon Press, Oxford, 1988, 526 pp.
[16] R.M. Aseeva, G.E. Zaikov "*Combustion of polymer materials*", Karl Hanser Verlag, Munchen, 1986, 389 pp.
[17] A.A. Popov, N.A. Rapoport, G.E. Zaikov "*Oxidation of stressed polymers*", Gordon and Breach, New York, 1991, 336 pp.
[18] A.F. Bochkov, G.E. Zaikov, V.A. Afanasiev "*Carbohydrates*", VSP Science Press VB, Zeist-Utrecht, 1991, 154 pp.
[19] V.A. Afanasiev, G.E. Zaikov "*Physical methods in chemistry*", Nova Science Publ., New York, 1992, 180 pp.
[20] I.N. Todorov, G.E. Zaikov, I A. Degterev "*Bioactive compounds: biotransformation and biological action*", Nova Science Publ., New York, 1993, 292 pp.
[21] V.L. Roubajlo, S.A. Maslov, G.E. Zaikov "*Liquid phase oxidation of unsaturated compounds*", Nova Science Publ., New York, 1993, 294 pp.
[22] A.L. Iordanskii, T.E. Rudakova, G.E. Zaikov "*Interaction of polymers with bioactive and corrosive media*", VSP International Publ., Utrecht, 1994, 298 pp.
[23] "*Degradation and stabilization of polymers. Theory and practice* ", Ed. by G.E. Zaikov, Nova Science Publ., New York, 1995, 238 pp.
[24] A.Ya. Polishchuk, G.E. Zaikov "*Multicomponent transport in polymer systems*", Gordon and Breach, New York, 1996, 231 pp.
[25] K.S. Minsker, A.A. Berlin "*Fast reaction processes*", Gordon and Breach, New York, 1996, 364 pp.
[26] E.Ya. Davydov, A.P. Vorotnikov, G.B. Pariyskii, G.E. Zaikov "*Kinetic pecularities of solid phase reactions*", John Willey and Sons, Chichester (UK), 1998, 150 pp.
[27] J.N. Aneli, L.M. Khananashvili, G.E. Zaikov "*Structuring and conductivity of polymer composites*", Nova Science Publ., New York, 1998, 326 pp.
[28] K.Z. Gumargalieva, G.E. Zaikov "*Biodegradation and biodeterioration of polymers. Kinetical aspects* ", Nova Science Publ., New York, 1998, 210 pp.
[29] S.K. Rakovsky, G.E. Zaikov "*Kinetics and mechanism of ozone reactions with organic and polymeric compounds in liquid phase*", Nova Science Publ., New York, 1998, 345 p.
[30] S.M. Lomakin, G.E. Zaikov "*Ecological aspects of polymer flame retardancy*", VSP International Publ., Utrecht, 1999, 158 pp.
[31] K.S. Minsker, G.E. Zaikov "*Chemistry of chlorine-containing polymers: synthesis, degradation, stabilization*", Nova Science Publ., New York, 2000, 198 pp.
[32] A. Jimenez, G.E. Zaikov "*Polymer analysis and degradation*", Nova Science Publ., New York, 2000, 287 pp.
[33] G.V. Kozlov, G.E. Zaikov "*Fractal analysis of polymers*", Nova Science Publ., New York, 2001, 244 pp.

[34] G.E. Zaikov, A.L. Buchachenko, V.B. Ivanov "*Aging of polymers, polymer blends and polymer composites*", Nova Science Publ., New York, 2002, Vol. 1, 258 pp.

[35] G.E. Zaikov, A.L. Buchachenko, V.B. Ivanov "*Aging of polymers, polymer blends and polymer composites*", Nova Science Publ., New York, 2002, Vol. 2, 253 pp.

[36] G.E. Zaikov, A.L. Buchachenko, V.B. Ivanov "*Polymer aging at the cutting adge*", Nova Science Publ., New York, 2002, 176 pp.

[37] S.A. Semenov, K.Z. Gumargalieva, G.E. Zaikov "*Biodegradation and durability of materials under the effect of microorganisms*", VSP International Science Publ., Utrecht, 2003, 199 pp.

[38] G.E. Zaikov, "*Burning, destruction and stabilization of polymers*", Scientific Basis and Technology Publishing House, St. Petersburg, 2008, 422 p. (in Rus.).

[39] A.K. Mikitaev, G.V. Kozlov, G.E. Zaikov, "*Polymer nanocomposites. A variety of structural forms and applications,* Nauka (Science) Publishing House, Moscoe, 2009, 279 p. (in Rus.).

[40] A.K. Mikitaev, G.V. Kozlov, G.E. Zaikov, "*Fractal analysis of the gas transport in polymers*, Nauka (Science) Publishing House, Moscow,, 2009, 199 p. (in Rus.).

[41] S.K. Rakovsky, G.E. Zaikov "*Interaction of ozone with chemical compounds. New frontiers* ", Rapra Technology, London, 2009, 486 pp.

[42] R.P. Naumova, *Conversion study of caprolactam by bacteria*, Ed. By N.A. Plate, Kazan State Technological University Publishing House, Kazan, 1964. - P.67-69 (in Rus.).

[43] R.P. Naumova, N.G. Zakharova, S.J. Zakharova, Degradation of synthetic lactam and ε-amino acids by microorganisms, cleaning industrial waste water. *Microbiological methods of dealing with environmental pollution*, Ed. By N.A. Plate, Nauka (Science) Publishing House (Pushchino Branch), Pushchino-town, 1979. - P.70-72 (in Rus.).

[44] K. Kato, T. Fukumura, Bacterial breakdown of ε-caprolactam, *Chem. and Industr. Journal,* 1962. - N 23. - P.1146.

[45] S. Kinoshita, E. Kobayashi, H. Okada, Degradation of ε-caprolactam by Achromobacter guttatus KF 71, *Journal of Fermentation Technology,* 1973, Vol. 51, P. 719-725.

[46] T. Uemura, Autolytic enzyme assocated with cell walls of Bacillus subtilis, *Journal of Biol. Chem.*, 1966, Vol. 241. - N 15. - P.3462 - 3467.

[47] A. Roy, *Destruction of caprolactam and hexamethylenediamine bacterial groups Bac. subtilis-mesentericus*, Thesis, Institute of Colloid and Water Chemistry, USSR Academy of Sciences, Kiev, 1975. – 162 pp. (in Rus.).

[48] A. Roy, The destruction of caprolactam microorganisms in continuous cultivation. *Scientific fundamentals of water treatment*, Ed. By N.A. Plate, Naukova Dumka (Publishing House of Ukrain Academy of Sciences), Kiev, 1976. - Issue 2. - P.152-156 (in Rus.).

[49] I.I. Shevtsova, *The destruction of microorganisms caprolactam*, Bulletin of Kiev State University, Division of Biology, Kiev, 1969. - № 11. - P.149-152 (in Rus.).

[50] T. Fucumura, Bacterial Breakdown of ε-caprolactam and its cyclic oligomers, Plant and cell. *Physiology Journal*, 1966. - Vol.7. - N1. - P. 93-104.

[51] R.P. Naumova, I.S. Belov, The transformation of aminocaproic acid in bacterial destruction of caprolactam, Biochemistry journal, 1968, № 33, p.946 (in Rus.).

[52] Biodamage: Textbook. *Manual for biological faculties of universities,* Ed. by V.D. Ilyicheva, High School Publishing House, Moscow, 1987, 352 pp. (in Rus.).

[53] E.L. Pekhtasheva, *Topical issues of biological damage*, Nauka (Science) Publishing House, Moscow, 1983, 265 pp. (in Rus.).

[54] A.N. Neverov, *Actual problems of biological damage and protection of materials, components and structures*, Science, Scientific Council for biological damage of the USSR Publishing House, Moscow, 1989, 256 pp. (in Rus.).

[55] A.N. Neverov, *Biodamage and methods for assessing the biological stability of materials,* Scientific Council for biological damage of the USSR Publishing House, Moscow, 1988, 140 pp. (in Rus.).

[56] A.N. Neverov, *Biodamage, methods of protection*, The Scientific Council for biological damage of the USSR Publishing House, Poltava, 1985, 182 pp. (in Rus.).

[57] I.G. Kanev, *Biological damage of industrial materials*, Nauka (Science) Publishing House, Leningrad, 1984, 232 pp. (in Rus.).

[58] *Biological problems of environmental materials*, Ed. by A.N. Neverov, Scientific Council on the biological damage of Sciences Publishing House, Penza, 1995, 108 pp. (in Rus.).

[59] *Methods for determining the biological stability of materials*, Ed. by N.A. Plate, Science Council of biological damage of the USSR Publishing House, Moscow, 1979, 202 pp. (in Rus.).

[60] Microorganisms and lower plants - *the destroyers of materials and products,* Ed. by N.A. Plate, Nauka (Science) Publishing House, Moscow, 1979, 225 pp. (in Rus.).

[61] *First Conference on Biological Damage*, Ed. by N.A. Plate, Nauka (Science) Publishing House, Moscow, 1978, 226 pp. (in Rus.).

[62] Problems of biological damage to materials. *Environmental aspects* Ed. by A.N. Neverov, Scientific Council of biological damage of the USSR Publishing House, Moscow, 1988, 124 pp. (in Rus.).

[63] *Environmental problems of biodegradation of industrial, construction materials and industrial wastes*, Ed. by A.N. Neverov, Scientific Council on Problems of biological damage Publishing House, Penza, 2000, 192 pp. (in Rus.).

[64] *Fourth Conference on Biological Damage*, Ed. by N.A. Plate, Scientific Council of biological damage of the USSR Publishing House, Nizhnii Novgorod-city, 1991, 99 pp. (in Rus.).

[65] E.I. Andreyuk, V. Bilan, E.Z. Koval, *Microbial corrosion and its agents,* Naukova Dumka, (Publishing House of Ukrain Academy of Sciences), Kiev, 1980, 258 pp. (in Rus.).

[66] R. Blagnik, B. *Newly, Microbiological corrosion*, Khimiya (Chemistry) Publishing House, Moscow, 1965, 222 pp. (in Rus.).

[67] Y.T. Shkut, A.F. Kostylev, *Histology and microbiology skins*, Light Industry Publishing House, Moscow, 1980, 150 pp. (in Rus.).

[68] V. Martha, K.S. Kutukova, Sh.D. Moszkowski, *Microbiology skins,* Gizlegprom Publishing House, Leningrad, 1936, 318 pp. (in Rus.).

In: Organic Chemistry, Biochemistry, Biotechnology ... ISBN: 978-1-62081-155-9
Editors: G. Zaikov, F. Pudel and G. Spychalski © 2013 Nova Science Publishers, Inc.

Chapter XVI

Biodamages and Protection of Wood

***Elena L. Pekhtasheva**, *Anatoly N. Neverov*,**
***Irina G. Shishkina and Gennady E. Zaikov*#**
G.V. Plekhanov Russian Economic University,Moscow, Russia
#N.M Emanuel Institute of Biochemical Physics,
Russian Academy of Sciences, Moscow, Russia

Abstract

The paper is collecting the information about biodegradation and stabilization of wood against of different fungi and bacteria actions. The contributors are discussing the biocide application for stabilization of wood.

Keywords: Biodamages, wood, protection, biodegradation, stabilization, biocides, fungi, bacteria

Along with metals and silicate materials (concrete, brick), wood forms a group of the most widespread materials people are using. However, in contrast with metals and silicate materials which are inorganic substances, wood is the natural organic matter and the source of carbon nutrition for many living organisms. Organisms utilizing wood are the source of biodamages of commercial timber, wooden structures, furniture and other articles. Development of microorganisms on these objects poses a hazard to human health, because microorganisms that damage them may cause skin, allergic diseases and be the source of aggressive toxins.

Wood is the universal raw material for multiple products manufacture. It is used in transport, mining, coal, paper making and wood chemical industry. The building industry

* E-mail: pekhtashevael@mail.ru.

consumes about 50% of total annual wood production in the form of lumber, industrial wood, plywood, various slabs and other articles, elements and structures.

The wide application of wood is based on its advantages: easy processing, high strength, possible high durability in dry conditions, beautiful appearance, etc. The main disadvantages of wood are combustibility, high water-absorbing capacity, and damageability by insects [1-5].

Various sorts of wood have different chemical composition (based on cellulose), as well as structure, density, strength of fibers and other properties, which affect biosability of the material. The plant raw material contains the following chemical components:

- cellulose – 27.0% (pine tree), 39.4% (birch);
- hemicelluloses – 17.8 (pine tree), 26.5% (birch);
- lignin – 19.7% (pine tree), 19.7% (birch);
- ash substances – 0.2% (pine tree), 0.4% (birch).

Chemical composition of absolutely dry wood includes the following elements: C – 49.5%; H – 6.3%; O + N – 44.2%, as well as mineral substances, mostly calcium salts in amounts of 0.2-1.7%.

Coniferous and leaf wood has the annual concentric rings, clearly observed due to formation of the spring wood, less dense and darker, and the autumn wood.

The main components of the wood macrostructure are the pitch or the heart (for the heartwood) – the central dark colored part, and the sapwood – light colored part neighboring the bark, the cortex. In contrast with leaf wood, coniferous wood has simpler structure and resin channels.

The pitch represents a loose tissue consisting of cells with thin, weakly bound walls. In the sawn wood, this is the weakest part of the stem, frequently crumbling and easily rotting.

The heart is the inner part of the stem forming, while the tree is growing, when the oldest inner part of the stem dies, conductive and storing system stops functioning, and cells pack. Water motion in these cells stops, therefore, the heart wood is strong and biostable. Some sorts of wood, e.g. birch or maple, have no heart.

Sapwood is formed by younger wood surrounding the heart. Water with dissolved nutrients flows by living cells of growing tree sapwood. The sapwood is highly humid and low biostable.

The wood consists of plant cells of various shapes, which perform particular functions in the tree life: tracheids – vessels representing the elements of the conductive system; parenchyma, which is the place of deposition of reserve nutrients (starch, fatty oils, sugars, etc.); libriform fibers implementing mechanical function.

Plant cells of coniferous and leaf wood are different. Coniferous wood is by 90-95% composed of tracheids. Spring tracheids are thin-walled water-conducting elements with wide cavities. Autumn tracheids represent thick-walled formations with less number of pores and cavities. They implement mechanical and storing functions. In coniferous wood, parenchyma gives 4-10% of total mass. These cells are rather small, with thin walls and multiple thinnings. They intercommunicate with other cells via pores.

The leaf wood structure is more complicated. Broad-leaved trees have harder wood. Mechanical strength of the leaf wood is provided by special cells – libriform fibers and fibrous tracheids. The birch wood has 65 to 70 vol.% of libriform fibers and fibrous tracheids. Parenchyma cells represent the storage tissue. In the oak wood, the parenchyma tissue gives 10-12%, which is 1.5-2-fold higher than in coniferous wood.

It is known that as compared with coniferous trees, the wood of many broad-leaved trees has lower natural biostability. The higher natural biostability, the lower additional chemical protection is required. The wood is divided into the following groups: biologically stable, medium stable, low stable and unstable [3].

- Biologically stable wood species are: yew, juniper, pine tree, ash tree, larch and oak heart;
- Medium stable are – spruce, cedar, fir tree, larch sapwood, birch heart;
- Low stable – elm, maple, birch and oak sapwood;
- Unstable – aspen, basswood, alder.

The main biodamaging agents for wood are microscopic fungi developing on it (wood inhabiting) and insects. In the midlatitudes, fungi give about 90% of all wood biodamages. Wood is generally biodamaged, because fungi and insects consume cellulose, lignin and other wood components as nutrients. As compared with fungi and insects which cause a direct damage of the wood fibers, bacteria induce lesser indirect damage.

In the damaged wood, fungi often form community with nitrogen-fixing bacteria, which enrich wood with accessible nitrogen. In contrast with bacteria, fungi easily penetrate into internal cavities of vessels and tracheids, water or air filled, due to the ability to form a branched hypha network. Bacteria mostly form colonies in parenchyma cells and inside the intracellular pores, where they dissolve primary cell wall membranes due to the action of cellulases, hemicellulases and pectinases enzymes. Destroying pore membranes in the sapwood, bacteria promote vide its aeration and, thus, stimulate fungus growth.

Curious Facts

When fungi propagate on the wood, its humidity increases. This happens at the account of water excretion, which is one of the final metabolism products. For instance, Serpula lacrymans fungus excretes 139 l of water per each destroyed 1 m^3 of wood; Coniophora puteana fungus during growth increases the initial wood humidity from 6.75 to 30-64%. Material humidity increase caused by fungi forms conditions for their invasion by new, more moisture-loving species of microorganisms.

The features of fungi metabolism, their demand in sources of carbon and energy have a very significant effect on the sequence of vegetable substrates colonization. For instance, fungi consuming easily accessible carbohydrates, sugars and hemicelluloses are first occurring on dead wooden substrates. This group includes forms capable of rapid occupation of space. They are characterized by quick spore germination in the presence of an appropriate substrate and accelerated mycelium growth. They are mucoraceous molds, *Penicillium, Aspergillus* genus fungi, etc.

While easily accessible nutrition sources are exhausted, the group of "sugar" fungi is replaced by slower propagating cellulose-degrading fungi: *Chaetomium sp.*, *Humicola grisea*, *Mycogone nigra*. Cellulose-degrading fungi of *Trichoderma* and *Fusarium* genus can develop in parallelwith "sugar" fungi or leave them behind. The latter are replaced by slowly propagating cellulose-degrading fungi and then by fungi, which degrade lignin. Lignin-degrading fungi begin developing, when easily accessible carbohydrates have already been exhausted by the primary inhabitants. This group includes slowly growing basidiomycetes, which complete cellulose degradation.

The fungi inducing wood biodamages are divided into three main groups:

- Surface mold fungi (moldy);
- Wood-staining;
- Wood-destroying:
    1. fungi inducing a soft homogeneous dote;
    2. fungi inducing brown rot of the degradative type;
    3. fungi inducing white rot of the corrosive type.

Surface mold fungi mostly colonize wet logs, lumber and various wood contaminations. Mold bloom formation is one of the initial signs testifying violation of storage or operation conditions of wood or articles from it. Surface molds usually destroy parenchyma tissues of the sapwood. Trichoderma, Cladosporium *and* Penicillium *genus fungi induce greenish staining, and* Aspergillus *and* Alternaria *genus fungi induce black spots.*

Wood-staining fungi are frequently developed on the wood at decelerated drying. They affect lumber, structures, wooden package, and stain wood in various colors.

One of the widespread staining is blue (blue stain fungi); yellow, orange, brown and other staining kinds are also observed. Wood-staining fungi have much in common with surface molds. Along with mold fungi, they are first colonizing wood by saprophytes, mostly consume wood nutrient reserves and, to a lesser extent, its structural elements, on which mechanical properties of wood depend. In contrast with molt fungi, wood-staining ones penetrate deep in the sapwood and induce deep staining by a pigment from hyphae and metabolites excreted by mycelium.

The highest damage to wood is caused by wood-destroying fungi, which are Scrpula, Coriolus, Fomitopsis, Serpula, Chaetomium, etc. genus fungi. Wood-destroying fungi destroy the structural components – cell walls of the wood. They inflate the living wood, humid and wet woody materials and articles from them. There are species among them, which better consume cellulose, and species consuming all three: cellulose, lignin and hemicellulose.

Among wood damages induced by wood-destroying fungi, by color and pattern three types of rot are outlined: white, brown and soft (moderate). White rot fungi primarily degrade lignin of the wood, leaving cellulose and hard wood areas untouched.

Curious Facts

In the 17th-18th centuries, Amati, Guarneri and Stradivarius in North Italy, in Cremona – the three dynasties of violin makers, won the lasting worldwide fame. The cherished dream of every violin player was to own a beautiful piece created by someone of those from Cremona. Antonio Stradivarius is especially famous till now. The most famous largest musicians play on his instruments.

In many countries, and in Italy itself, people tried to crack the Cremona secrets. But nobody succeeded yet in making a violin, which would possess the same melodiousness, as the instruments by these maestros, which are extant.

Joseph Nagivari, the American biophysicist, has published a report, which interested not only specialists, but also the wide circle of music lovers. As to his opinion, the wood of these instruments is not so good, e.g. for joiners articles.

In several violins by the famous maestros from Cremona, Nagivari has found traces of microscopic fungus activity. He has found documents indicating that the wood was delivered to Cremona via mountain rafting by the Po river, on which bank the town is located. Nagivari has obtained wood contaminated by microscopic fungi of the same species, as the old violins are, and has made several violins. The experienced musicians agreed that these violins assonated the old ones from Cremona.

The scientist thinks that, apparently, fungi destroy resinous matter of the wood, it becomes softer and better vibrates.

Nagivary says that in the recent one and half hundred years, violin-makers make the same mistake, making violins from wood, which is too dry and seasoned and has not external and internal defects.

Brown rot fungi prefer cellulose, leaving brown areas of wood after damaging, which easily powder when touched. Brown rot can be the reason for destruction of wooden structures and facilities.

There is one more type of wood biodamages – carpet rot: at this type of rotting, small focuses of low wood damages form a many-colored carpet image of alternating grey, bluish, brown and yellow spots. Wood infected by rot of this type is easily penetrable for water and at incessant rain may become completely wet.

By destructive impact on the wood, rots are classified as destructive and corrosive, which types differ by the action mechanism, which is clearly observed in the analysis of lignin/cellulose ration dynamics. For instance, this ratio typically increases for pine tree infected by destructive rot. Wood infection by corrosive rotting causes a relative increase of cellulose content that affects the lignin/cellulose ratio. For larch, this index does not virtually change, and for pine tree it tends to decrease.

The wood infection by rot induces cellulose, lignin, pentosans, resins and fats degradation. The content of substances soluble in water and mild alkaline increases during wood destruction, which is the obvious indicator of degradation. With respect to the infection stage, wood density decreases by 32-39% against sound wood.

Wood biodamages are usually combined with aging caused by weather, mechanical and other operational loads. Fibers of the wood surface layer are loosened and woolliness increases (maceration) under the effect of regular moistening, temperature change, sunlight and other factors. Moisture and dust are accumulated here, and conditions favorable for fungus spore development are formed.

Wood infected by rot absorbs water much easier. Existing cracks expand due to wood shrinkage. Water frozen in the cracks increases destruction; shatters and drops off of the sapwood occur. The wood heart becomes more exposed.

Along with microscopic fungi, the wood is also damaged by insects. Although the general loss from insects is lower, in some cases and regions, beetles, especially termites, are so dangerous that application of special protective measures is mandatory.

The following defects are unacceptable:

- External rotten decay, when damaged wood easily disintegrates and powders;
- Sapwood rot shaped as spots and stripes over 1/3 of square;
- Tobacco knots, in which the wood is fully or partly replaced by a loose snuff or hoary colored mass, or doted ones with a soft rot occupying over 1/3 of the knot cut;
- Worm holes – channels and holes (shallow, deep and through) made by insects in the wood; only surface, below 3 mm deep ones are allowed in the case of closing.

Wood Protection against Biodamages

Protection shall be complex, including measures on prophylaxis of biodamages, preventing moistening of wood, rationally using its natural protective features, selecting appropriate woods, developing optimal design solutions and, finally, applying chemical protection means – biocides (called antiseptics in the wood protection practice) [6-18].

Chemical protection of wood against biodamages is applied in cases of the most heavy-duty operation of wood, for example, at permanent or regular contact with the soil, humid atmosphere and water. About 5-10% of wood, totally used for industrial purposes and in construction should be processed by biocides. As a result, the lifetime of articles and structures increases by several times. For instance, correct application of antiseptics can increase the life time of common wooden houses from 15 to 50 years, on average.

In Russia, the loss of wood in the timber industry and woodworking, at the stages of harvesting, storage, transportation and processing reaches 60%, whereas in developed countries (e.g. Finland or Sweden) this does not exceed several percents.

In Russian practice, antiseptics are applied to exported sawn goods, round timber not protected at all.

Antiseptics for the wood protection shall:

- protect wood against wood-staining and mold fungi and conserve it for, at least, two months;
- be stable to atmospheric impacts and not stain the wood;
- be able to sufficiently penetrate into the wood to the least depth of 0.5 mm;
- be well soluble in water;
- be personnel-friendly;
- be stable during storage, not evaporate with toxicant release.

Water-soluble antiseptics represent solid friable powders applied as aqueous solutions, suspensions and pastes.

The main antiseptic of this group are: sodium (potassium) dichromate usually applied in combination with copper sulfate; sodium and ammonium fluorides and silicofluorides; sodium pentafluorophenoxide. Selkur preparation based on copper oxide and chromium anhydride is widely used. Impregnation of wood with this compound forms a layer of acid copper chromate on the fiber surface. This layer is almost water insoluble and thus provides protection antiseptic against washing off during wood operation.

The disadvantage of water-soluble antiseptics is the necessity of additional wood drying procedure. Furthermore, wood drying results in shrinkage and crack formation.

There are many methods of protective wood impregnation by antiseptics, which can be divided into two groups: impregnation by wood submerging into the solution; surface spraying of antiseptic or brushing.

When wood is submerged into liquid antiseptic, the latter penetrates into pores, cracks and chaps. To increase antiseptic absorption, dense wood with small pores is preliminarily needled on a special machine. Impregnation is performed in baths filled with the antiseptic solution, to which wood is exposed. Deeper impregnation and better protection is provided by hot solvent application prior to cold antiseptic solution. Heating expands and partly removes air present in cracks and pore of wood. Vacuum formed in pores at air contraction caused by cooling down provides deeper penetration of the antiseptic solution into the pores.

Wood protection against biodamages is usually combined with protection against combustion. For this purpose, flame retardants are added to the antiseptic solution.

Some woody materials, such as plywood, fiber boards and chip boards, etc. are also treated by antiseptics applied by adding a biocide to a binder of scale wood sheets or by impregnating the prepared material [19-21]

References

[1] *Protection of wood and cellulose materials against biological damage*, Ed. by V.P. Karlivan, The Scientific Council for biological damage of the USSR Publishing House, Riga, 1989, 267 pp. (in Rus.).

[2] V.P. Salovarova, Y.P. Kozlov, *Ecological and biotechnological bases conversion plant substrates*, Energy Publishing House, Moscow, 2006, 544 pp. (in Rus.).

[3] Biodamage: Textbook. *Manual for biological faculties of universities*, Ed. by V.D. Ilyicheva, High School Publishing House, Moscow, 1987, 352 pp. (in Rus.).

[4] V.D. Il'ichev, B.V. Bocharov, M.V. Gorlenko, *Environmental framework for the protection of biological damage*,Nauka (Science) Publishing House, Moscow, 1985, 264 pp. (in Rus.).

[5] A.A. Imshenetsky, *Microbiology of cellulose*, USSR Academy of Sciences Publishing House, Moscow, 1953, 439 pp. (in Rus.) E.L. Pekhtasheva, *Topical issues of biological damage*, Nauka (Science) Publishing House, Moscow, 1983, 265 pp. (in Rus.).

[6] A.N. Neverov, *Actual problems of biological damage and protection of materials, components and structures*, Science, Scientific Council for biological damage of the USSR Publishing House, Moscow, 1989, 256 pp. (in Rus.).

[7] A.N. Neverov, *Biodamage and methods for assessing the biological stability of materials*, Scientific Council for biological damage of the USSR Publishing House, Moscow, 1988, 140 pp. (in Rus.).

[8] A.N. Neverov, *Biodamage, methods of protection*, The Scientific Council for biological damage of the USSR Publishing House, Poltava, 1985, 182 pp. (in Rus.).

[9] I.G. Kanev, *Biological damage of industrial materials*, Nauka (Science) Publishing House, Leningrad, 1984, 232 pp. (in Rus.).

[10] *Biological problems of environmental materials*, Ed. by A.N. Neverov, Scientific Council on the biological damage of Sciences Publishing House, Penza, 1995, 108 pp. (in Rus.).

[11] *Methods for determining the biological stability of materials,* Ed. by N.A. Plate, Science Council of biological damage of the USSR Publishing House, Moscow, 1979, 202 pp. (in Rus.).

[12] Microorganisms and lower plants - the destroyers of materials and products, Ed. by N.A. Plate, Nauka (Science) Publishing House, Moscow, 1979, 225 pp. (in Rus.).

[13] *First Conference on Biological Damage*, Ed. by N.A. Plate, Nauka (Science) Publishing House, Moscow, 1978, 226 pp. (in Rus.).

[14] Problems of biological damage to materials. *Environmental aspects* Ed. by A.N. Neverov, Scientific Council of biological damage of the USSR Publishing House, Moscow, 1988, 124 pp. (in Rus.).

[15] *Environmental problems of biodegradation of industrial, construction materials and industrial wastes*, Ed. by A.N. Neverov, Scientific Council on Problems of biological damage Publishing House, Penza, 2000, 192 pp. (in Rus.).

[16] *Fourth Conference on Biological Damage*, Ed. by N.A. Plate, Scientific Council of biological damage of the USSR Publishing House, Nizhnii Novgorod-city, 1991, 99 pp. (in Rus.).

[17] "*Degradation and stabilization of polymers. Theory and practice* ", Ed. by G.E. Zaikov, Nova Science Publ., New York, 1995, 238 pp.

[18] A.Ya. Polishchuk, G.E. Zaikov "*Multicomponent transport in polymer systems*", Gordon and Breach, New York, 1996, 231 pp.

[19] G.E. Zaikov, A.L. Buchachenko, V.B. Ivanov "*Aging of polymers, polymer blends and polymer composites*", Nova Science Publ., New York, 2002, Vol. 1, 258 pp.

[20] G.E. Zaikov, A.L. Buchachenko, V.B. Ivanov "*Aging of polymers, polymer blends and polymer composites*", Nova Science Publ., New York, 2002, Vol. 2, 253 pp.

[21] G.E. Zaikov, A.L. Buchachenko, V.B. Ivanov "*Polymer aging at the cutting adge*", Nova Science Publ., New York, 2002, 176 pp.

In: Organic Chemistry, Biochemistry, Biotechnology ... ISBN: 978-1-62081-155-9
Editors: G. Zaikov, F. Pudel and G. Spychalski © 2013 Nova Science Publishers, Inc.

Chapter XVII

Biomass Logistics: The Key Challenge of Minimizing Supply Costs

Ina Ehrhardt*, Holger Seidel and Christian Blobner
Fraunhofer Institute for Factory Operation and Automation IFF,
Magdeburg, Germany

Abstract

More than 53 % of the renewable energy consumed in Germany is generated from biomass, predominantly wood, and the trend is upward. Research has focused on the utilization of woody biomass and neglected the issues and challenges confronting the actors in the biomass value added and supply chain. Biomass suppliers and consumers are usually unaware of the share of total supply costs incurred by complex logistics and indirect costs. Hence, holistic logistics concepts and bases for planning are needed to supply biomass cost effectively and sustainably as cross-company value added.

Keywords: Logistics, biomass, direct and indirect costs, supply chain management, value added chain, renewable energy, cost cutting

Background and Objectives

Supported by such instruments as the German Renewable Energy Act (EEG) and other European and national support programs, the utilization of renewable energies is increasingly growing in importance. Biomass already supplies around 70% of the energy obtained from renewable energy sources.[1] Whereas biomass (solid biofuels, liquid and gaseous fuels and

* E-mail: Best4VarioUse@fraunhofer.de.
[1] German Federal Ministry of the Environment, Nature Conservation and Nuclear Safety BMU, Erneuerbare Energien in Zahlen - nationale und internationale Entwicklung, June 2010, p. 14.

biowaste) only accounts for 19.8 % and thus a small part of the renewable power produced, it is the source of 91.5% of the renewable heat produced.[2] The raw material wood constitutes the most significant proportion of solid biofuels used to supply heat.

Woody biomass generally denotes thinnings, forest wood residues and wood from landscape conservation (roadside vegetation, natural copses, prunings, driftwood). However, energy wood from plantations, industrial and other wood waste wood are also woody biomass.

The utilization of woody biomass is generating competition for this resource now grown scarce. Utilizing woody waste and residues from agriculture, forestry and landscape conservation represents one solution to this conflict. In the Best4VarioUse project, the Fraunhofer IFF is presently researching issues surrounding the supply and utilization of wood biomass residues. Together with fifteen partners from Germany and the Spanish region of Valencia, the Fraunhofer IFF is researching the limitations on organizing economically and ecologically efficient biomass supply chains and the implications for concepts for regional utilization.

Biogenic wastes and residues from biotope and landscape conservation, agriculture and forestry contain characteristic amounts of woody biomass dependent on their provenance and composition. (See Table 1 for a summary of forestry, agriculture and landscape conservation actions that produce woody biomass waste and residues.) So far, these residues have remained largely unutilized. Efficiently utilizing raw materials hitherto classified as waste in energy value chains will necessitate implementing and, where necessary, modifying combined methods and technologies so that the material flows are economically and ecologically efficient. The low local biomass yields and wide dispersion of consumers render the great complexity of the logistics of the biomass supply a key challenge.

Table 1. Actions producing woody biomass waste and residues

Domain	Action				
Forest	Clearance after damage	Regular thinning/ selective felling[3]	Micro-development	Firebreaks	Cultivation of succession lands[4]
Landscape conservation	Tree care	Preservation of open landscapes	Roadside vegetation		
Agriculture	Hedgerow care	Deshrubbing	Pruning/ tree care	Cultivation of abandoned land[5]	Clearance
Other	Construction site clearance	Traffic safety[6]			

Efficient planning and organization of the logistical structures and exact knowledge of the direct and indirect costs incurred in the supply chain are the basis for the future

[2] BMU, Erneuerbare Energien in Zahlen, p. 13.
[3] Including bird cherry.
[4] Including land under power lines.
[5] Including access road construction.
[6] For example, near buildings, tracks, roads and other transportation routes.

development of the utilization of biomass residues and wastes. Residues and wastes will only become a true alternative once they can be supplied at reasonable prices and in proper qualities. Moreover, the cost drivers in the supply chain will have to be identified in order to cuts costs.

Factors Influencing the Biomass Supply Chain

Methods and tools of supply chain management (SCM) provide the basic principles and means to analyze logistical issues related to the value added chain and the organization of the biomass supply chain. These tools raise companies' awareness of the overall operation and thus increase cross-company value added. SCM revolves around cross-company coordination of material and information flows throughout the entire value added process, from raw material production to individual stages of processing up through the end consumer, with the goal of organizing the overall operation so that both time and cost have been optimized. Logistics interfaces are crucial to a supply chain's smooth operation. Hence the goal must be to optimize the interfaces by eliminating unnecessary activities and unnecessary resource consumption at system interfaces in order to optimize costs.

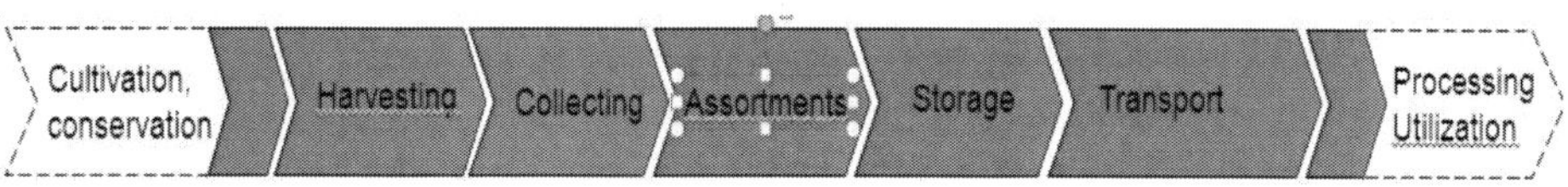

Figure 1. Biomass supply chain.

A systematic analysis is needed to optimize biomass value added ecologically and economically. First, the operational stages of a typical supply chain are analyzed. Then, select factors that influence auf direct and indirect supply and logistics costs are mapped.

Biomass logistics operations subsume every stage from harvesting to final delivery to processing. Since upstream and preparatory forest stand planning and tending operations or operations in a wood processing plant are not considered direct elements, they are only analyzed tangentially. The planning, control, execution, monitoring and billing of all biomass harvesting, storage, processing and transport operations and even related information flows are also part of biomass logistics. The typical overall supply chain from the "forest to the factory" encompasses the individual operations of harvesting, extraction, assortment formation, storage and transport (see Figure 1).

Numerous factors, which can be categorized as external or internal influences, determine the amount of the costs incurred by supplying biomass. They affect indirect or direct costs related to operations or raw material (see Table 2). For instance, hauling distances and transfer points (forest roads and temporary storage facilities) affect transport costs. The number of competing providers in a region or the prices of competing fossil fuels also affect the price of biomass, which must not only cover supply costs but also furnish the actors involved in the operation a profit margin.[7]

Table 2. Morphology of factors influencing costs incurred

Factors	
Influencing direct costs	Influencing indirect costs
Internal	External
Operations-related	Raw material-related

Biomass Supply Costs

Generally, biomass supply or logistics costs are incurred by the operations to supply the raw material from its site of production to consumers. Five different types of logistics costs have been distinguished.[8]

1. System and control costs, i.e. costs incurred by the organization, planning and monitoring of the material flow and by production program planning, material planning, order processing and production control.
2. Inventory costs, i.e. costs incurred by maintaining and financing stocks.
3. Storage costs, i.e. costs incurred by maintaining storage capacities and storage and retrieval operations.
4. Transport costs, i.e. costs incurred by internal and external plant transport.
5. Handling costs, i.e. costs incurred by packaging and picking partial shipments from a total ordered shipment.

Supply chain costs must be identified for analysis and optimization. Part of the costs, e.g. harvesting costs, can be measured directly, thus eliminating the need for detailed research since they are relatively obvious. Indirect costs, e.g. administrative or communication costs, are more difficult to identify. A study conducted by the Fraunhofer IFF in Magdeburg revealed that indirect supply costs are usually not transparent, thus neither quantifiable nor qualifiable and therefore frequently not entered into costing and pricing. Consequently, biomass providers are aware in the planning phase of the direct (procurement) costs of services such as harvesting, skidding or transport but not the total costs that offset the "value" of the particular biomass supplied. Thus, marketing and assortment decisions are usually based only on the direct supply costs and the obtainable market price and factor in no other criteria and variables. Especially in Germany, the large number of actors and differing contracts contained in the value added chain significantly affect the amount of the indirect costs.

[7] Schuster, K.: Überbetriebliche Aufbringung und Vermarktung von Brennhackschnitzeln. Landtechnische Schriften No. 179, Österreichisches Kuratiorium für Landtechnick, Vienna, 1991.

[8] Schulte, C. Logistik: Wege zur Optimierung der Supply Chain. Vahlen: Munich, 2005.

The Fraunhofer IFF has determined that indirect costs account for 9-20 % of the total costs. One study identified certain factors that most frequently affect indirect costs in the overall biomass supply chain:

- information and communications technology (internal and external, i.e. partners'),
- the type, duration and frequency of information exchange and
- internal and external communication.

Clearly, factoring in transport and harvesting costs does not suffice to cut total costs.

Field tests in the Best4VarioUse project corroborated the data gathered in interviews and studies. Several different scenarios were developed for harvesting, storage, processing and transport and methods were developed to ascertain the direct and indirect costs. Among other things, morphological approaches were employed to define actions (see table 1) and select the technology to be tested.

The actions selected for the field tests were typical agricultural, forestry and landscape conservation actions, e.g. the creation of networks of tending and skidding trails, which have not delivered any commercially developable yield of biomass but are necessary actions nonetheless, e.g. in forestry, with potentially utilizable biomass yields.

The indirect costs, e.g. for supply planning and coordination, turned out to constitute a significant portion of the biomass supply costs, but are not normally enter into costing. The smaller the biomass yields, the more considerable indirect costs. Oversized special equipment used by service providers, which is not justifiable for a cost effective cross-company supply chain, is another cost driver.

Interaction and the Information Flow Are Crucial

Forestry operations, wood processing industries and the other actors involved in the chain of operations plan and control their activities based on internal, normally functional, criteria and their areas of responsibility (e.g. purchasing, logistics or marketing). The criteria are largely a function of the size of the service companies', e.g. forestry service or transport providers, and their equipment and services. The different quantities ordered by and the quality requirements of differently sized consumers, from single households to medium-sized combustion plants, present another challenge.

Moreover, the many different actors are interconnected differently in different scenarios and the widest variety of client-contractor relationships exist, as the following two scenarios illustrate. In one scenario, the owner of a private forest sells the biomass from it and contracts different service providers to harvest and transport it. A consumer transports it and has a service provider chip it to the requisite quality before burning it. In another scenario, the forest owner has employees harvest and chip the biomass with a leased chipper and then has it transported to the consumer by a transport provider. Both scenarios rely on different actors and therefore also generate different direct and, above all, indirect costs.

In addition to technical problems at the interfaces between the individual actors in the biomass supply chain, deficits in organization and IT especially impede organizing biomass value added cost effectively and ecologically sustainably. Interactions between and factors influencing the actors and their impact on the cost effectiveness and efficiency of internal and cross company operations are often overlooked.

These operations often drive indirect costs in particular. Normally, many small orders are accepted and filled in the biomass value added chain. Indirect costs are incurred by both small and large orders. Although the costs for small orders are lower than for large orders, they increase exponentially when the total quantity ordered in both small and large orders add up to be identical. Thus, concluding larger contracts, i.e. consolidating orders, can significantly cut indirect costs and thus also total costs. Many different contractors offer their services and different terms and conditions (price, type of contract and quality). The multitude of interfaces and lack of advanced information technologies often cause communication losses and thus longer wait and search times or duplicated work. Information is often gathered or transmitted incompletely and absent standards generates additional costs and cause information losses at interfaces. Therefore, good interaction among the actors and a coordinated information flow are crucial to reducing indirect costs.

Planning Is Crucial

From the site of production to the consumer, the biomass supply chain is characterized by many spatially distributed, time-variable biomass sources and many spatially distributed providers and consumers (see Figure 2). A multitude of actors (forest owners, forestry operations, private forestry contractors, carriers, transport companies, combustion plant operators, etc.) must collaborate in the strongly customer-oriented supply chain. This diverse and complex network of partners collaborates in different operations and exchanges information through a multitude of interfaces.

Biomass's different properties (e.g. shape, weight and volume), different methods of harvesting widely varying quantities and assortments and different means of processing at individual locations constitute other specific challenges when planning operations. The quantity and composition of the biomass may vary greatly depending on the type, scope and location of the forestry, agricultural and landscape conservation actions. The quality and quantity of the biomass from landscape conservation actions or forest thinning may also vary greatly. Consumer requirements vary just as greatly. Small and medium-sized combustion plants require chipped wood measuring between 8.5 cm and 25 cm in size (length).

This raises the question: What quantity of which biomass is to be harvested with what equipment and at what cost for which consumer(s)? Thus, the focus of cost optimization shifts to strategic and operative planning. Strategic planning must include an early, integrated analysis of the site of production and the biomass consumers. Identifying the location and plant cover of lands, the infrastructure, distances to consumers and their requirements, and the time and scope of actions is essential.

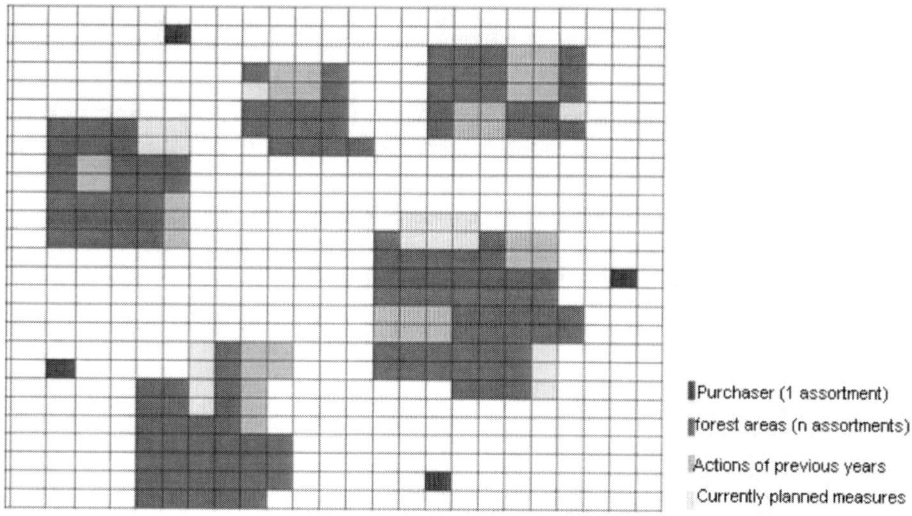

Figure 2. Spatially distributed, time-variable biomass sources and spatially distributed providers and consumers.

Operatively, planning for the right equipment in the right place at the right time is crucial. This especially affects direct supply costs, e.g. incurred by the failure to fully utilize expensive special equipment. In addition, harvesting and processing equipment greatly affect biomass's quality.

The aforementioned field tests in the Best4VarioUse project corroborated these statements. The practical tests explicitly compared different methods of harvesting and processing. Motor-manual harvesting with chainsaws, brushcutters or similar special equipment often proves to be superior, especially economically. Large and special equipment is often inappropriate for the material's small dimensions or the actions themselves. On the other hand, motor-manual harvesting is extremely flexible and thus expedites processing.

Optimizing the Overall Operation

Optimal strategic and operative planning is indispensible to optimizing operations and costs. Marketing decisions can only be made when costs and sale prices have been compared throughout the entire biomass supply chain, factoring in different scenarios and alternative consumers. Continuous modification of the equipment used ensures that the biomass is supplied as efficiently as possible. This entails determining whether special or universal equipment should be invested in or existing machinery can be modified with less expense. Information on the site of use, operating conditions and restrictions and the quality of the biomass are important for such decisions. Additionally, this information is an indispensible criterion for selecting methods. Pooling different harvesting, processing or transport equipment is an expedient cost cutting strategy.

Taking advantage of potentials for collaborations to systematically combine interfaces and capabilities is another means to optimize total operational costs. Any commercial risks need to be assessed beforehand. Ultimately, the range of services can usually be expanded.

Thinking outside of the box is indispensible to taking this initial step toward a value added network. Individual operations and activities must be examined critically in order to be able to draft contracts expediently and calculate costs realistically. The correlation between one's own and one's network partners' success must always be borne in mind.

Consistent use of advanced ICT can significantly cut costs in the overall operation and is instrumental to the successful implementation of integrated planning based on sound foundations and proper tools with standardized interfaces accessible to all of the biomass supply chain's actors.

When these strategies and approaches are rigorously followed, the optimized planning and control of the biomass supply chain cuts indirect and direct costs significantly and sustainably.

Conclusion and Outlook

Logistics costs include far more than just harvesting and transport costs. Optimizing the value added chain ecologically and economically entails factoring in not only the direct but also the indirect supply costs of the entire biomass supply chain. The Fraunhofer IFF in Magdeburg is currently pursuing research to formulate recommendations for actions that will minimize direct costs, among other things, by applying optimal planning and control. In addition, potentials are being identified to cut indirect costs by optimizing the operations throughout the entire biomass supply chain. This will constitute a substantial step toward optimizing the biomass value added chain ecologically and economically.

References

Aretz A., Hirschl B. Biomassepotenziale in Deutschland-Übersicht maßgeblicher Studienergebnisse und Gegenüberstellung der Methoden. *DENDROM Discussion Paper* No.1, March 2007.

Dreier T. *Ganzheitliche Systemanalyse und Potenziale biogener Kraftstoffe*. E und M, Energie-und-Management-Verlag-Ges.: Herrsching, 2000.

Dietz, H.-U.; Nick, L.; Ehrhardt, I.; Urbanke, B.; Hauck, B. *Standardisierung in der Forstlogistik – Notwendigkeit, Wirkung und Chancen*. In Thees O. und Lemm R. (Eds.): Management zukunftsfähige Waldnutzung Grundlagen, Methoden und Instrumente. vdf Hochschulverlag AG: Zurich, 2009.

Fritsche et al. *Stoffstromanalyse zur nachhaltigen energetischen Nutzung von Biomasse*,. Darmstadt, May 2004.

German Federal Ministry of the Environment, Nature Conservation and Nuclear Safety BMU: *Erneuerbare Energien in Zahlen - nationale und internationale Entwicklung*, Berlin, June 2010.

Kaltschmitt, M.; Merten, D. et al. *Energie aus Biomasse: Grundlagen, Techniken und Verfahren*. Springer Verlag: Berlin, 2009.

Kreis Siegen-Wittgenstein. Landschaftsplanung in Naturschutzgebieten (NSG). www.siegen-wittgenstein.de/standard/page.sys/details/eintrag_id= 790/content_id=588/349.htm. July 30, 2010.

Kuhn A. und Hellingrath B. *Supply Chain Management - Optimierte Zusammenarbeit in der Wertschöpfungskette*. Springer Verlag: Berlin, 2002.

Murach, D.; Knur, L. und Schultze, M. *DENDROM – Zukunftsrohstoff Dendromasse Systemische Analyse, Leitbilder und Szenarien für die nachhaltige energetische und stoffliche Verwertung von Dendromasse aus Wald- und Agrarholz*. Eberswalde, November 2008.

Schneider St. Entwicklung eines (teil-)automatisierten IT-Prozesses zur Erzeugung digitaler Wegedaten unter Berücksichtigung landwirtschaftlicher Aspekte. *Unpublished Diplom Thesis submitted to the Department of Business Information Systems,* School of Computer Science, Otto von Guericke University Magdeburg. 2010.

Schulte, C.: *Logistik: Wege zur Optimierung der Supply Chain*. Vahlen: Munich, 2005.

Schuster, K.: *Überbetriebliche Aufbringung und Vermarktung von Brennhackschnitzeln*. Landtechnische Schriften No. 179, Österreichisches Kuratiorium für Landtechnick, Vienna, 1991.

Ropohl G. *Eine Systemtheorie der Technik. Zur Grundlegung der allgemeinen Technologie*. Hanser Verlag: Munich, 1979.

Holzzentralblatt. *Integrierte Logistikketten für die Holzbranche*. DRW-Verlag Weinbrenner GmbH + Co. KG Holz-Zentralblatt. July 15, 2005.

Landesforstverwaltung NRW. *Clusterstudie Forst and Holz*. http://www.forst.nrw.de/nutzung/cluster/cluster.htm. 2003.

In: Organic Chemistry, Biochemistry, Biotechnology ... ISBN: 978-1-62081-155-9
Editors: G. Zaikov, F. Pudel and G. Spychalski © 2013 Nova Science Publishers, Inc.

Chapter XVIII

Ecological and Economic Optimisation of Biogas Plants

Georg Häring[1], Matthias Sonnleitner[1,*], Wilfried Zörner[1], and Vic Hanby[2]
[1]Centre of Excellence for Renewable Energy Research,
Ingolstadt University of Applied Sciences, Esplanade, Ingolstadt
[2]Institute of Energy and Sustainable Development,
De Montfort University, Leicester (UK)

Abstract

This paper reviews the analysis of 10 biogas plants in Bavaria (Germany). It is shown that improvements in ecology and economy can be achieved with simple, low cost and quickly implementable solutions. The contribution presents the results of a research project and shows tangible suggestions to improve biogas plants in ecology and economy.

Introduction

2008 was a difficult year for the German biogas sector. After a dramatic price increase for energy crops in summer 2007, the following autumn saw the beginning of discussions about the amendment of the German Renewable Energies Act (EEG). Every few years, this law, which regulates the compensation paid for electricity from renewable primary products, is reviewed.

* Tel.: 0841-9348 649, Fax: 0841-9348 99649. E-Mail: matthias.sonnleitner@haw-ingolstadt.de, Internet: www.haw-ingolstadt.de.

The amendment in 2004 preceded a boom in the years 2004 to 2006 followed by a phase of stagnation in the biogas sector in the years 2007 and 2008. The demand for new production facilities fell to almost zero and the biogas plant operators suffered from the dramatic price increase for energy crops. Some farmers were even forced to close their biogas production facilities as a reaction to the lack of profit potential. The amendment to the EEG was introduced in 2009, and the economic situation of the biogas sector has since improved.

Incomplete and premature planning, low-quality construction work and a lack of operators' skill are still the major problems agricultural biogas plants are faced with (Sonnleitner and Zoerner 2007). These problems directly interfere with the economic and ecological situations (Sonnleitner and Zoerner 2008). Agricultural biogas production plants can become an economic risk for biogas plant operators, i.e. the farmers. Furthermore, these facilities can produce notable methane emissions to the atmosphere which prevent a long-term environmentally friendly energy supply. For these reasons, a research project supported by the *German Federal Ministry for the Environment* was carried out, in order to improve the ecological and economic performance of agricultural biogas plants by generating guidelines for biogas operators, planners and manufacturers with tangible suggestions for improvements. In order to achieve this, 10 biogas plants in Bavaria (Germany) were thoroughly analysed and evaluated.

The first stage of the project was to produce a widespread data analysis. The analysis considered a number of areas for comparison, such as the measurement of the parasitic electric energy or the detection of methane leaks. The operational log of the 10 biogas plants was also analysed with respect to a number of categories, including feedstock, down times, labour times and maintenance.

The next stage of the project was to evaluate the generated data. For this purpose, key performance indicators were used to carry out a benchmark and error analysis.

The increased awareness of the problems biogas plants are faced with finally led to the development of optimisation concepts. Therefore, 10 approaches were carried out to improve the ecology and economy of the plants.

Analysis of Current Situation

The first part of this paper focuses on particular results of a widespread data analysis. The electric capacity of the selected plants is between 30 and 560 kWel. These 10 plants are marked as BY1...BY10 in table 1.

Desulphurisation

Upon closer inspection of the 10 biogas plants, it was found that the specific desulphurisation energy consumption via air injection varies between 0.02 and 0.42 $(kWh_{el}/d)/(Nm^3/h)$ (figure 1). There is a high specific desulphurisation energy consumption in the smaller biogas plants. This cannot be attributed to a higher air injection rate, but can be traced back to an inadequate dimensioning and design of the air injection system.

The biological desulphurisation of the biogas plants BY1 and BY6 is carried out via air compressors which show significantly higher energy consumption than normal desulphurisation air blowers.

Due to the broad range of fluctuation, there is a high potential for the reduction of the desulphurisation energy consumption.

Table 1. Selected biogas plants in Bavaria (Germany)

NR.	kWel	CHP-Unit	Feedstock	Start of Operation
BY1	30	Pilot Injection	RRM + Slurry	1999
BY2	100	Gas		2002
BY3	175	Gas		2004
BY4	250	Gas		2005
BY5	190	Gas		2005
BY6	320	Pilot Injection		2001
BY7	380	Gas		2005
BY8	380	Gas	RRM	2006
BY9	526	Gas	RRM + Slurry	2006
BY10	560	Gas		2002

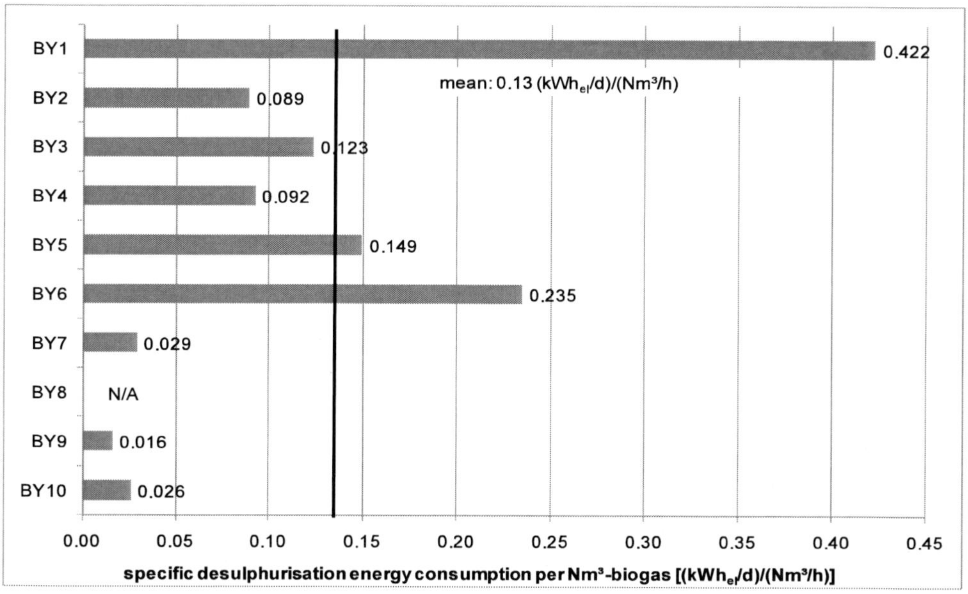

Figure 1. Specific desulphurisation electric energy consumption.

Biogas Leaks

Figure 2 shows the identified biogas leaks of the investigated biogas plants. These leaks were identified using a leakage detector, designed to detect methane. To identify methane leaks, the main focus was on all components which are responsible for the conduction of biogas or are directly connected to active substrate.

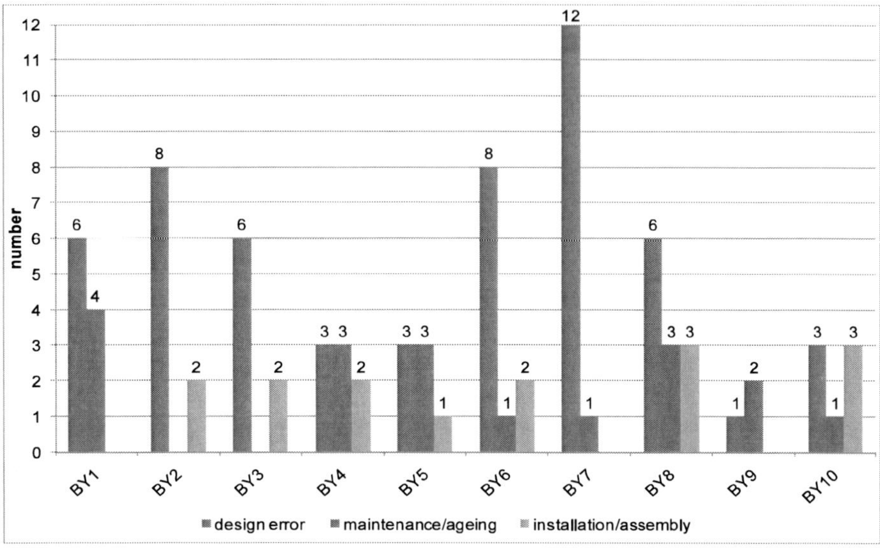

Figure 2. Identified and categorised biogas leaks.

Furthermore, the identified biogas leaks are divided into 3 categories: methane emissions caused by design errors (e.g. open overflows or emergency opening in digesters), deficiencies due to lack of maintenance/ due to ageing (e.g. leaks at gasholders), mistakes in assembly or installation (e.g. incorrectly installed inspection windows).

In this investigation, most of the detected biogas leaks are caused by design errors. Manufacturers of biogas plants can achieve a significant reduction in methane emissions by avoiding those design errors.

Parasitic Electric Energy

Figure 3 shows the parasitic electric energy of the 10 biogas plants. Thus, the electric energy consumption of each biogas plant is measured and compared to the gross electricity output. Within this analysis, the parasitic electric energy is subdivided into 2 categories: the demand for the biogas production and the demand for the CHP-Unit.

The biogas production process contains a number of electric consumers such as the feeding system, agitators, pumps, gas processing and circulation pumps for heating the digester. The CHP-Unit contains electric consumers such as a gas compressor, a primary / secondary pump, an emergency cooler, a charge air cooler and room ventilator.

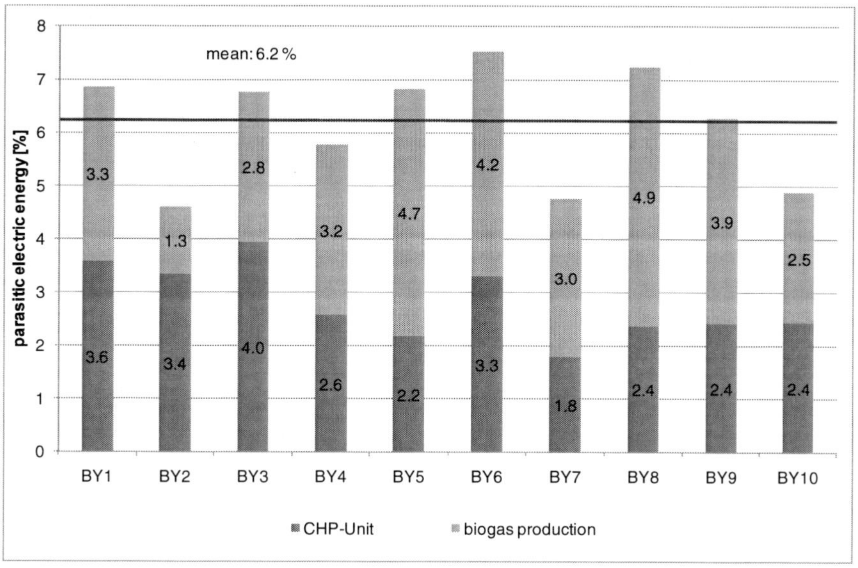

Figure 3. Parasitic electric energy.

The parasitic electric energy of the 10 biogas plants varies from 4.7 % (BY2) to 7.5 % (BY6). The proportion of the mean CHP-Unit electric energy consumption is 2.8 % whereas 3.3 % is the average needed to run the components of the biogas production.

Development of Optimisation Concepts

Based on the analysis of the current situation, a benchmark is carried out according to the following criteria:

- economic potential for optimisation,
- ecological potential for optimisation,
- potential for realisation.

Table 2. Evaluation criterions

	Rating		
	Low	medium	high
Economic Potential for Optimis action	low additional income	medium additional income	high additional income
Ecological Potential for Optimis action	low reduction of emissions	medium reduction of emissions	high reduction of emissions
Potential for Realisation	implemented in most plants	implemented in many plants	rarely implemented in most plants
Points	1	2	3
Overall Score	3	4-5	6-9

Table 3. Evaluation with regard to potential for optimisation

Areas of investigation	Economic Potential for Optimis action	Ecological Potential for Optimis action	Potential for Realisation	Overall Score
Substrate Supply				
Distance to Farmland	Low	Low	Low	3
Required Farmland	Low	Low	Low	3
Substrate-Origin: Owned or Purchased	Low	Low	Low	3
Loading of Substrate				
Distance Silo – Feeding System	High	Low	Medium	6
Time for Loading the Feeding System	High	Low	Medium	6
Biogas Production				
Fermentation Process	High	Low	Medium	6
Used Substrates	Low	Medium	Low	4
Biology	Low	Low	Medium	4
Gas Processing				
Desulphurisation Efficiency	Medium	Low	Medium	5
Dehumidification Efficiency	Medium	Low	Medium	5
Biogas Utilisation				
CHP-Unit	High	High	High	9
Heat Utilisation	High	High	High	9
Measurement and Control Technology				
Measurement	Medium	Medium	Medium	6
Documentation	Medium	Low	Medium	5
CHP-Unit Usage Rate	High	Medium	Medium	7
Parasitic Electric Energy				
Total Parasitic Electric Energy	High	Medium	Medium	7
Feeding System	High	Low	Medium	6
Agitator	High	Low	Medium	6
Pump	Medium	Low	Medium	5
Desulphurisation	High	Low	High	7
Dehumidification	High	Low	Medium	6
Power Supply	High	Low	Medium	6
Peak Loads	Medium	Low	High	6
Methane Emissions				
Biogas Leaks	Medium	High	High	8
Remaining Biogas Potential	High	High	Medium	8

This evaluation finally leads to the development of optimisation concepts with tangible suggestions for improvements.

The rating is carried out according to the classification and allocation of points as shown in table 2.

Topics which are rated with 6...9 points are promising targets with a high potential for improvement. The result of the evaluation matrix is shown in table 3.

Based on the evaluation matrix, the following 10 approaches were carried out to improve the plants` ecology and economy:

- Shortening the distance between silo and feeding system
- Selecting the optimum dehumidification method
- Improving substrate conversion by using highly efficient CHP-Units
- Improving substrate conversion by avoiding biogas leaks
- Reducing methane emissions and improving substrate conversion by covering residue storage tanks
- Lowering desulphurisation electric energy consumption via air injection
- Lowering feeding system electric energy consumption
- Improving heat utilisation by structured approach in planning
- Improving heat utilisation by implementing heat meters
- Improving utilisation of the CHP-Unit

In the course of this paper, two optimisation concepts are described thoroughly.

Reduction of Desulphurisation Electric Energy Consumption via Air Injection

The biological desulphurisation within the digester via air injection is considered to be an inexpensive, as well as a reliable and low-maintenance technique. With this method the hydrogen sulphide content can be reduced effectively.

The biogas plants` desulphurisation electric energy consumption (DEEC) can be calculated. A comparison of the various energy demands for injecting air into the digester shows that the energy demand cannot be attributed to plant-specific factors (e.g. injected air flow, growing layer for sulphur bacteria).

Higher specific DEEC can be easily identified at smaller plants, and can be traced back to an inadequate dimensioning and design of the air injection system (e.g. oversized air blowers, use of air compressors).

Figure 4 shows a formula that can be used to calculate the specific DEEC.

The DEEC should not exceed 0.05 (kWh_{el}/d)/(Nm^3/h).

$$\text{specific desulphurisation electric energy consumption}\left[\frac{kWh_{el}/d}{Nm^3/h}\right] = \frac{\text{el. power desulphurisation}\left[kW_{el}\right] \cdot \text{runtime}\left[h/d\right]}{\text{biogas flow}\left[Nm^3/h\right]}$$

Figure 4. Formula to calculate the specific desulphurization electric energy consumption.

Oversized Air Blowers

Figure 5 shows an oversized air blower with an electric energy consumption of 540 W_{el}. To limit the injected air flow to the required 70 l/min, the ball valve is almost closed.

So, this biogas plant with a biogas flow of 85 Nm^3/h consumes three times more electric energy (0.149 $(kWh_{el}/d)/(Nm^3/h)$) for desulphurisation than the target value of 0.05 $(kWh_{el}/d)/(Nm^3/h)$. Thus, about € 710 has to be paid for electricity if an electricity price of 0.15 €/kWh_{el} is assumed.

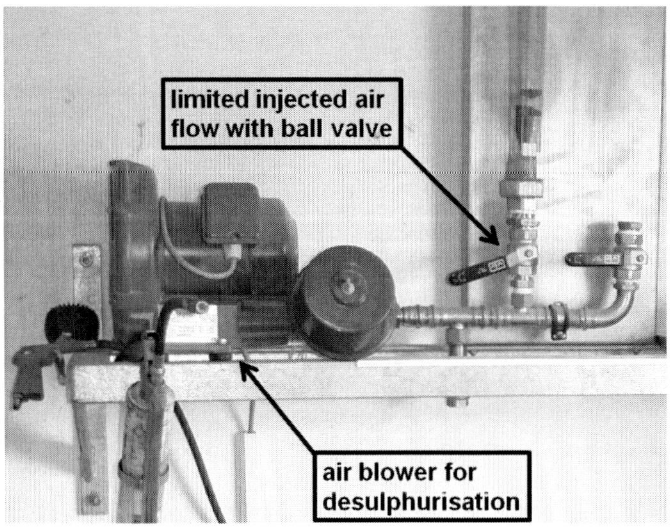

Figure 5. Oversized air blower for desulphurization.

To save on electric energy consumption, either the air blower can be replaced by an adequately dimensioned air blower or a timer can be used to limit the runtime of the oversized air blower.

For example, an air blower with an electric energy consumption of 85 kW_{el} can be used as an alternative. This device will also deliver the required air flow. Using the 85 kW_{el} device in the previously mentioned biogas plant will result in a specific DEEC of 0.024 $(kWh_{el}/d)/(Nm^3/h)$. That is equivalent to an electricity cost of € 112, which means a saving of about € 600 per year. In comparison to the annual saving, the initial investment is just € 350.

Besides the saving of electric energy and money also a reduction of emissions of about 2,500 kg CO_2-Eq/a (emissions of the German electricity grid: 630 g CO_2-Eq/kWh_{el}) is achieved. This is tantamount to the emissions of a car driving about 14,000 km.

Air Injection via Air Compressor

The injection of air is in some cases done via air compressors. Initially, this method seems to make sense, because air compressors are needed to run the pneumatic valves of pumping stations.

However, air compressors provide air at high pressure and this pressure must be lowered before the air is injected into the digester. This means a high percentage of waste energy.

The following example shows how much energy is wasted by using air compressors for desulphurisation.

For the biological desulphurisation of biogas at a flow rate of 115 Nm^3/h, an air compressor is used that injects 90 l/h. About 2/3 of the pressurised air is used for desulphurisation, as shown in figure 6 and figure 7. Every day 14 kWh_{el} is needed to provide the required air flow. This means a specific DEEC of 0.12 $(kWh_{el}/d)/(Nm^3/h)$. That is equivalent to a payment of € 767 for electricity per year, if an electricity price of 0.15 €/kWh_{el} is assumed.

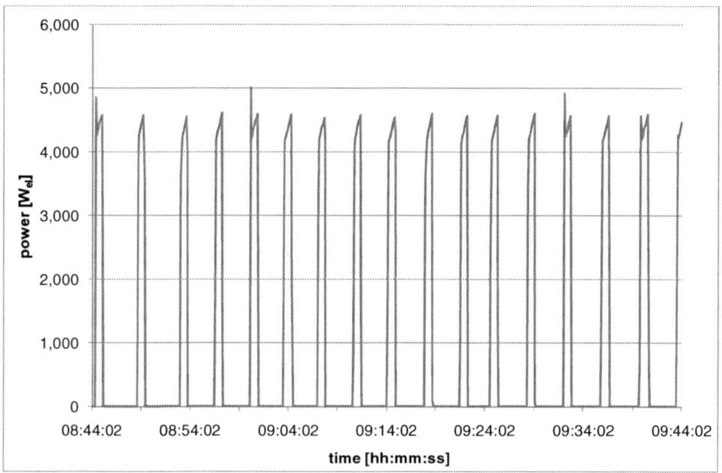

Figure 6. Power consumption of an air compressor running a desulphurisation and pneumatic valves of a pumping station.

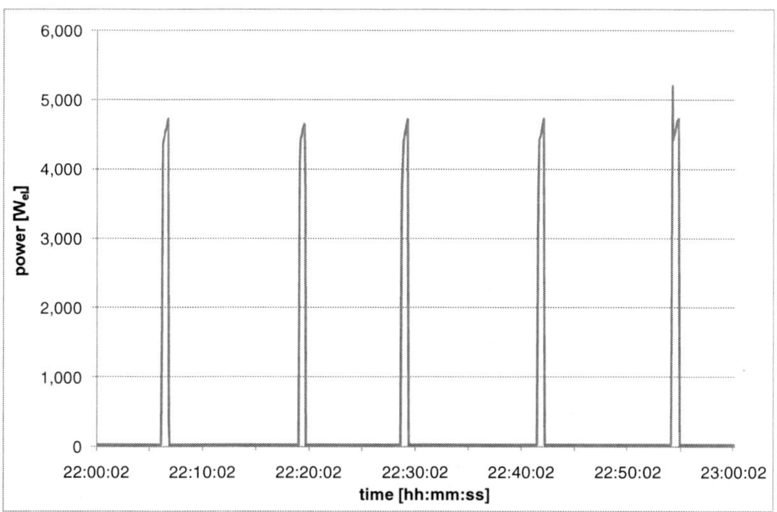

Figure 7. Power consumption of an air compressor without running a desulphurization.

If the air compressor was replaced with an air blower (electric energy demand: 95 kW_{el}; investment: € 350...400), the specific DEEC would be reduced to 0.02 $(kWh_{el}/d)/(Nm^3/h)$. The equivalent payment for electricity would be € 125, resulting in a saving of € 642 per year. Besides the saving of electric energy and money, a reduction of emissions of about 2,700 kg CO_2-Eq/a (emissions of the German electricity grid: 630 g CO_2-Eq/kWh_{el}) would also be achieved. This is tantamount to the emissions of a car driving about 15,000 km.

Table 4. Frequently identified biogas leaks

Biogas Leak		Recommended Action
Feeding screw		Piston feeding system and/or feeding below substrate level
Open overflow		Avoidance of open overflows in general
Wire rope grommet of a submersible motor mixer		Re-grease the grommet after each movement of the wire rope
Mounting agitator	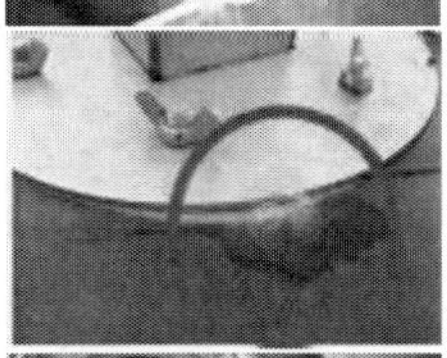	Tighten the bolts; assembly according to manufacturer (mistake in assembly)
Porthole	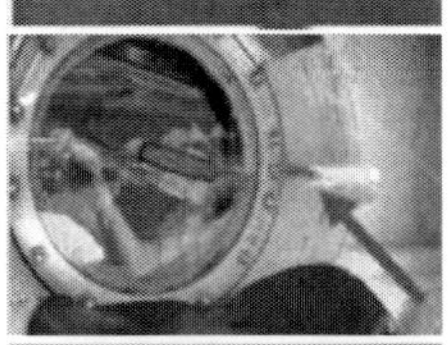	Fix gas backflow preventer valve and ball valve at provided place (mistake in assembly)
Gasholder		Sealing according to manufacturer
Flange connection		Assembly according to manufacturer (mistake in assembly: less screws)
Emergency opening		Replace sealing

Better Substrate Conversion via Reduction of Biogas Leaks

Diffuse emissions of methane are unnecessary losses that can be identified and avoided quite easily. In most cases, this can be done by the biogas plant operator himself.

As a precondition, a leakage detector, designed to detect methane, must be available. The investment for a leakage detector (explosion-proof) is about € 400...600. There is also the possibility of buying the leakage detector in collaboration with other plant operators to reduce the cost per biogas plant.

Biogas plants should be inspected once a month to detect biogas leaks. A higher frequency of inspections is recommended if plant modifications are carried out or biogas leaks are assumed (e.g. odour of biogas).

During different operational conditions, such as higher pressure inside the digester, more biogas leaks can occur than normal. Thus, inspections with a leakage detector should be done while the system is under different operating conditions.

Table 4 shows common biogas leaks.

The identified biogas leaks are divided into 3 categories: methane emissions caused by design errors, deficiencies due to lack of maintenance or ageing and mistakes in assembly or installation.

Biogas leaks caused by mistakes in assembly and installation (e.g. incorrectly installed inspection windows or flange connections) as well as deficiencies due to lack of maintenance or ageing (e.g. worn seals) can be fixed easily and do not infer high costs.

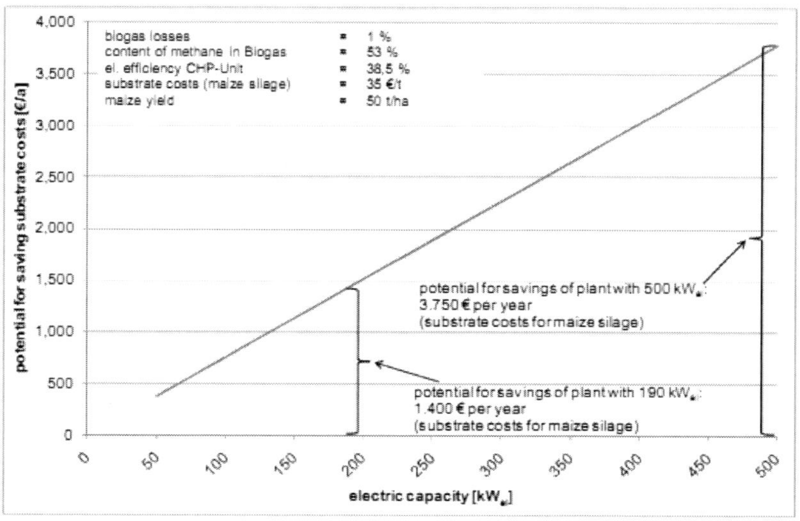

Figure 8. Potential for savings via reduction of biogas leaks.

It is important to note that the detection of biogas leaks is carried out in a potentially explosive atmosphere. Thus, precautions have to be taken, especially during maintenance and repairs.

Figure 8 shows the savings of substrate costs (maize silage) per year. The recovery of 1 % biogas losses with regard to the produced biogas has a high potential for reduced costs.

In the case of a 190 kW_{el} biogas plant, the following savings can be achieved:

- Reduced substrate costs (maize silage): 1,400 €/a,
- Reduced farmland usage (maize silage): 0.8 ha,
- Saved Substrate (maize silage): 41 t/a,
- Avoided methane emissions: 4,161 m^3/a,
- Reduced greenhouse gas emissions due to the reduction in methane emissions and saved substrate: 75,917 kg CO_2-Eq./a (emissions of ca. 7 persons per year).

A 500 kW_{el} biogas plant can achieve the following savings:

- Reduced substrate costs (maize silage): 3,750 €/a,
- Reduced farmland usage (maize silage): 2.1 ha,
- Saved Substrate (maize silage): 108 t/a,
- Avoided methane emissions: 10,950 m^3/a,
- Reduced greenhouse gas emissions due to the reduction of methane emissions and saved substrate: 199,785 kg CO2-Eq./a (emissions of ca. 17.5 persons per year).

Both examples prove that avoiding biogas leaks contributes significantly to the protection of climate and to more profitable businesses.

Conclusion

The analysis of 10 biogas plants in Bavaria (Germany) showed that improvements in ecology and economy can be achieved with simple, low cost and quickly implementable solutions.

Implementing the suggested solutions, both greenhouse gas emissions and current expenses can be reduced. Furthermore, the competition for farmland with the food production sector is diminished.

As expected, the biogas plant operator is an important factor for the profitability of a biogas plant. Thus, the following preconditions need to be fulfilled:

- Basic education regarding the operation of biogas plants,
- Working with attention to detail:
  - compliance with maintenance schedules,
  - complete operational log.

It should be noted that many deficiencies such as high parasitic electric energy and biogas leaks can be traced back to design errors. To avoid these faults in the future, the suggested solutions should be taken into consideration in the planning process.

References

Göbel, A.; Sonnleitner, M.; Zörner, W. Biogas: Chance oder Risiko für die Landwirtschaft? Erkenntnisse aus der Feldstudie „Biogasanlagen in Bayern“, In: *16. Symposium BIOENERGIE, Bad Staffelstein, November 2007.* Regensburg: OTTI – Bereich Erneuerbare Energien, pp. 230-235.

Sonnleitner, M. Und Zörner, W. (2008) Biogas: Chance oder Risiko für die Landwirtschaft – Update der Feldstudie „Biogasanlagen in Bayern“, In: *17. Symposium BIOENERGIE, Bad Staffelstein, November 2008.* Regenburg: OTTI – Bereich Erneuerbare Energien, pp. 176-181.

In: Organic Chemistry, Biochemistry, Biotechnology ... ISBN: 978-1-62081-155-9
Editors: G. Zaikov, F. Pudel and G. Spychalski © 2013 Nova Science Publishers, Inc.

Chapter XIX

Novel Polydiphenylamine-2-Carbonic Acid/Fe_3O_4 Magnetic Nanoparticles

I. S. Eremeev**, S. Zh. Ozkan and G. P. Karpacheva
Russian Academy of Sciences A.V. Topchiev
Institute of Petrochemical Synthesis, Moscow, Russia

Abstract

Magnetic nanoparticles based on Fe_3O_4 and polydiphenylamine-2-carbonic acid (PDPhACA) with core-shell structure, where Fe_3O_4 is the core and PDPhACA is the shell, were obtained for the first time. Formation of nanoparticles was proved by X-ray diffraction analysis and transmission electron microscopy. On X-ray images one can clearly identify diffraction peaks of Fe_3O_4 nanoparticles. Fe_3O_4/PDPhACA nanoparticles dimensions fall in the limits of $4<d<14$ nm. IR-spectroscopy results have shown that polymeric shell is formed by means of C-C joining into para-position of phenyl rings with respect to nitrogen. Magnetic properties of Fe_3O_4/PDPhACA nanoparticles were studied. It was found that obtained nanoparticles are superparamagnetic.

Keywords: Polydiphenylamine-2-carbonic acid; magnetic Fe_3O_4/PDPhACA nanoparticles; core-shell structure

Materials based on polyconjugated polymers and magnetic nanoparticles attract special interest due to their unique electric, magnetic and optical properties. Magnetic nanoparticles based on Fe_3O_4 and polydiphenylamine-2-carbonic acid (PDPhACA) were obtained for the first time [1]. Formation of Fe_3O_4/PDPhACA nanoparticles involves synthesis of Fe_3O_4 nanoparticles through a hydrolysis of mixture of chlorides of Fe (II) and (III) in relation 1:2 in ammonium hydroxide solution [2], monomer fixation on the surface of Fe_3O_4 nanoparticles with further in situ polymerization with the addition of ammonium persulfate. As a result nanoparticles with core-shell structure are formed, where Fe_3O_4 is the core and PDPhACA is

* E-mail: eremeev@ips.ac.ru.

the shell. The polymeric shell, that has positively charged groups in its structure, prevents the aggregation of nanoparticles.

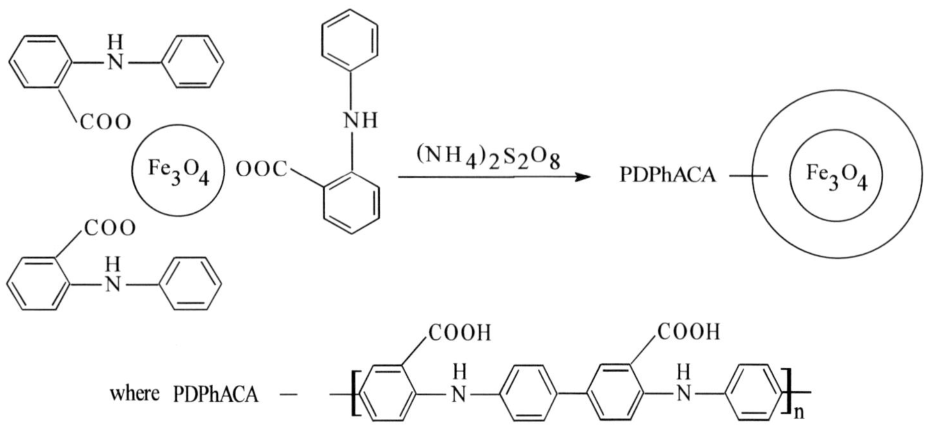

Structure of Fe_3O_4/PDPhACA nanoparticles was proved by XRD and TEM. On X-ray images one can clearly identify diffraction peaks of Fe_3O_4 nanoparticles (Figure 1). According to TEM data, Fe_3O_4/PDPhACA nanoparticles dimensions fall in the limits of $4<d<14$ nm (Figure 2).

IR-spectroscopy results have shown that polymeric shell is formed by means of C-C joining into para-position of phenyl rings with respect to nitrogen (Figure 3). Absorption bands near 884 and 800 cm^{-1} are due to nonplanar deformation vibrations $\delta_{C\text{-}H}$ of 1,2,4 and 1,4-substituted aromatic rings [3].

Investigation of magnetis properties of Fe_3O_4/PDPhACA nanoparticles at room temperature (Figure 4) has shown that the obtained nanomaterial is superparamagnetic (Ms = 33.5 emu/g, Hc = 76 Oe, Mr/Ms = 14.9%).

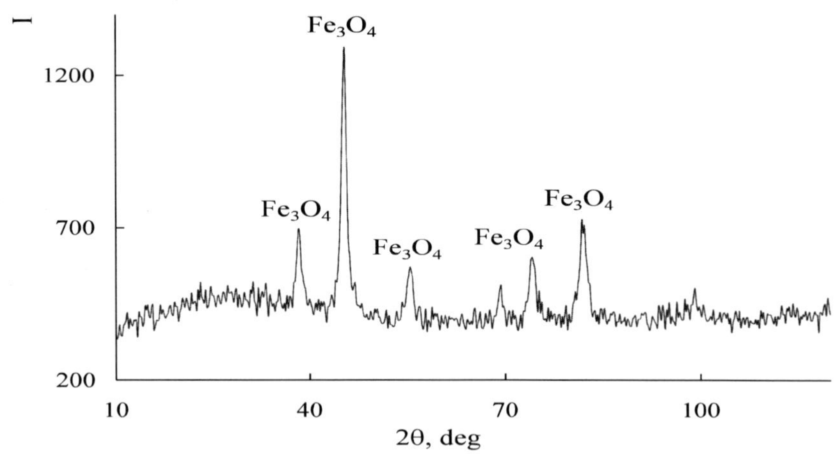

Figure 1. XRD pattern of Fe_3O_4/PDPhACA nanoparticles.

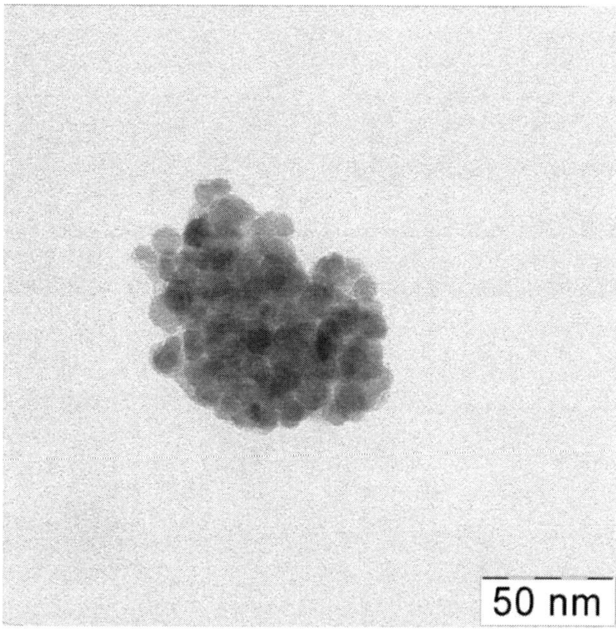

Figure 2. Microphotograph of Fe_3O_4/PDPhACA nanoparticles.

Superparamagnetic Fe_3O_4/PDPhACA nanoparticles can be used as magnetic components in aqueous and organic ferrofluids, self-stabilization of which will be provided by polymeric shell.

The work has been supported in part by the Russian Foundation for Basic Research, project 11-03-00560a.

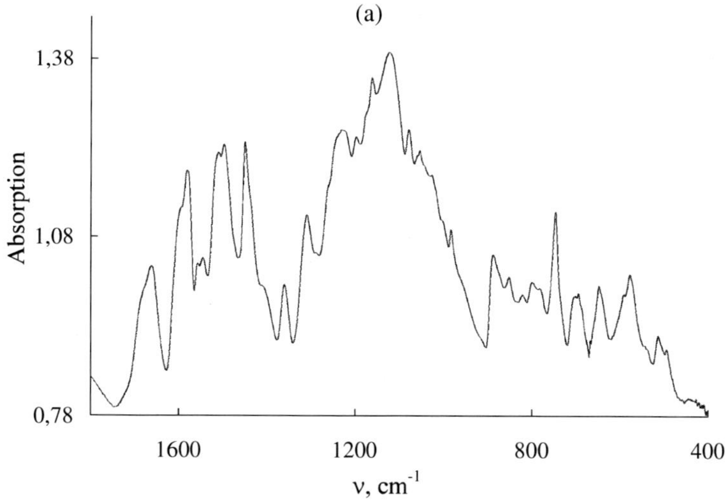

Figure 3. (Continued)

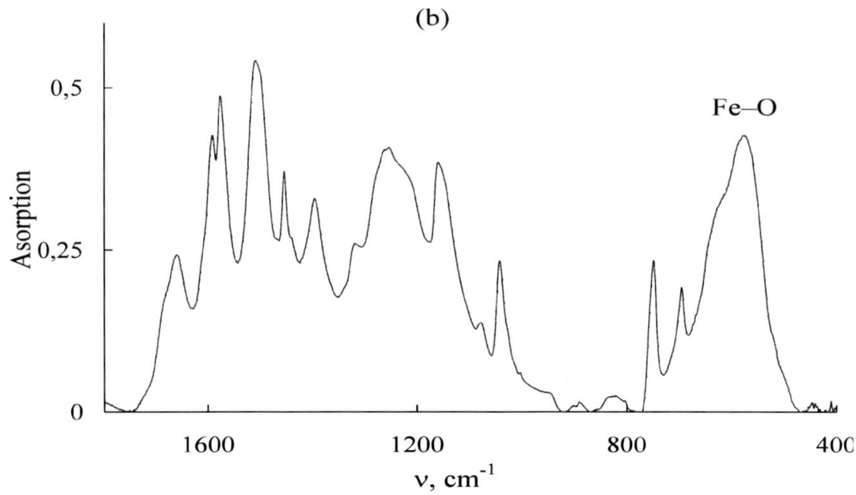

Figure 3. IR-spectra of PDPhACA (a) and Fe_3O_4/PDPhACA nanoparticles (b).

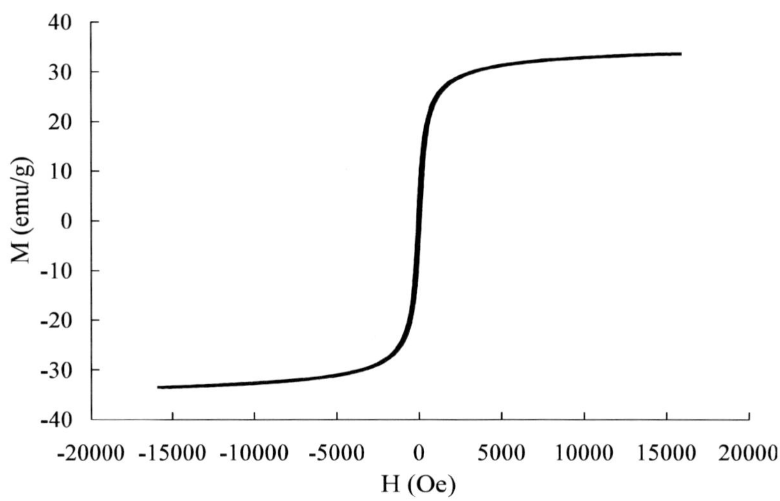

Figure 4. Magnetization as a function of applied magnetic field at room temperature.

References

[1] Karpacheva G.P., Ozkan S.Zh. / RF patent for the invention of "*Dispersed nanocomposite magnetic material and method of its obtaining*", № 2426188 from 10.08.2011.

[2] Massart R. // *IEEE Trans. Magn*. 1981. Mag-17. V. 2. P. 1247.

[3] Ozkan S.Zh., Bondarenko G.N., Karpacheva G.P. // *Polym. Sci. B*. 2010. V. 52. № 5. P. 263.

Index

D

E

F

G

H

I

J

K

N

O

P